London Mathematical Society Student Texts 78

Clifford Algebras: An Introduction

D. J. H. Garling

LONDON MATHEMATICAL SOCIETY STUDENT TEXTS

Managing Editor: Professor D. Benson,
Department of Mathematics, University of Aberdeen, UK

LONDON MATHEMATICAL SOCIETY STUDENT TEXTS 78

Clifford Algebras: An Introduction

D. J. H. GARLING
Emeritus Reader in Mathematical Analysis,
University of Cambridge, and
Fellow of St John's College, Cambridge

CAMBRIDGE
UNIVERSITY PRESS

CAMBRIDGE
UNIVERSITY PRESS

University Printing House, Cambridge CB2 8BS, United Kingdom

Published in the United States of America by Cambridge University Press, New York

Cambridge University Press is part of the University of Cambridge.

It furthers the University's mission by disseminating knowledge in the pursuit of education, learning and research at the highest international levels of excellence.

www.cambridge.org
Information on this title: www.cambridge.org/9781107422193

© D. J. H. Garling 2011

First published 2011

A catalogue record for this publication is available from the British Library

ISBN 978-1-107-09638-7 Hardback
ISBN 978-1-107-42219-3 Paperback

Cambridge University Press has no responsibility for the persistence or accuracy of URLs for external or third-party internet websites referred to in this publication, and does not guarantee that any content on such websites is, or will remain, accurate or appropriate.

Contents

Introduction

Clifford algebras find their use in many areas of mathematics: in differential analysis, where operators of Dirac type are used in proofs of the Atiyah-Singer index theorem, in harmonic analysis, where the Riesz transforms provide a higher-dimensional generalization of the Hilbert transform, in geometry, where spin groups illuminate the structure of the classical groups, and in mathematical physics, where Clifford algebras provide a setting for electromagnetic theory, spin 1/2 particles, and the Dirac operator in relativistic quantum mechanics. This book is intended as a straightforward introduction to Clifford algebras, without going on to study any of the above topics in detail (suggestions for further reading are made at the end). This means that it concentrates on the underlying structure of Clifford algebras, and this inevitably means that it approaches the subject algebraically.

The first part is concerned with the background from algebra that is required. The first chapter describes, without giving details, the necessary knowledge of groups and vector spaces that is needed. Any reader who is not familiar with this material should consult standard texts on algebra, such as Mac Lane and Birkhoff [MaB], Jacobson [Jac] or Cohn [Coh]. Otherwise, skim through it, to familiarize yourself with the notation and terminology that is used.

The second chapter deals with algebras, and modules over algebras. It turns out that the algebra $\mathbf{H}$ of quaternions has an important part to play in the theory of Clifford algebras, and fundamental properties of this algebra are developed here. It also turns out that the Clifford algebras that we study are isomorphic to the algebra of $\mathbf{D}$-endomorphisms of $\mathbf{D}^k$, where $\mathbf{D}$ is either the real field $\mathbf{R}$, the complex field $\mathbf{C}$ or the algebra $\mathbf{H}$; we develop the theory of modules over an algebra far enough to prove Wedderburn's theorem, which explains why this is the case.

Tensor products of various forms are an invaluable tool for constructing Clifford algebras. Many mathematicians are uncomfortable with tensor products; in Chapter 3 we provide a careful account of the multilinear
algebra that is needed. This involves finite-dimensional vector spaces,
where there is a powerful and effective duality theory, and we make
unashamed use of this duality to construct the spaces of tensor products
that we need.

The second part is the heart of this book. Clifford algebras are constructed, starting from a vector space equipped with a quadratic form.
Chapter 4 is concerned with quadratic forms on finite-dimensional real
vector spaces. The reader is probably familiar with the special case of
Euclidean space, where the quadratic form is positive definite. The general case is, perhaps surprisingly, considerably more complicated, and we
provide complete details. In particular, we prove the Cartan-Dieudonné
theorem, which shows that an isometry of a regular quadratic space is
the product of simple reflections.

In Chapter 5, we begin the study of Clifford algebras. This can be
done at different levels of generality. At one extreme, we could consider
Clifford algebras over an arbitrary field; these are important, for example in number theory, but this is too general for our purposes. At the
other extreme, we could restrict attention to Clifford algebras over the
complex field $\mathbf{C}$; this has the advantage that $\mathbf{C}$ is algebraically complete, which leads to considerable simplifications, but in the process it
removes many interesting ideas from consideration. In fact, we shall consider Clifford algebras over the real field; this provides enough generality
to consider many important ideas, while at the same time provides an
appropriate setting for differential analysis, harmonic analysis and mathematical physics. We shall see that the complex field $\mathbf{C}$ is an example
of such a Clifford algebra; one of its salient features is the conjugation
involution. Universal Clifford algebras also admit such an involutory automorphism, the *principal involution*. This leads to a $\mathbf{Z}_2$ grading of such
algebras; they are *super-algebras*. This is one of the fundamental features
of the structure of Clifford algebras. But they are more complicated than
that; besides the principal involution there are two important involutory
anti-automorphisms.

Much of the charm of Clifford algebras lies in the fact that they have
interesting concrete representations, and in Chapters 6 and 7 we calculate many of these. In particular, we use arguments involving Clifford
algebras to prove Frobenius' theorem, that the only finite-dimensional
real division algebras are the real field $\mathbf{R}$, the complex field $\mathbf{C}$ and the

algebra **H** of quaternions. The calculations involve dimensions up to 5; we also establish a partial periodicity of order 4, and Cartan's periodicity theorem, with periodicity of order 8, which enable all Clifford algebras to be calculated. As a result of these calculations, we find that every Clifford algebra is either isomorphic to a full matrix algebra over **R**, **C** or **H**, or is isomorphic to the direct sum of two copies of one of these. This is a consequence of Wedderburn's theorem: a finite-dimensional simple real algebra is isomorphic to a full matrix algebra over a finite-dimensional real division algebra. Matrices act on vector spaces; at the end of Chapter 7 we introduce *spinor spaces*, which are vector spaces on which a Clifford algebra acts irreducibly.

A large part of mathematics is concerned with symmetry, and the orthogonal group and special orthogonal group describe the linear symmetries of regular quadratic spaces. An important feature of Clifford algebras is that the group of invertible elements of a Clifford algebra contains a subgroup, the *spin group*, which provides a double cover of the corresponding special orthogonal group. In Chapter 8, spin groups are defined, and their basic properties are proved. Spin groups, and their actions, are calculated for spaces up to dimension 4, and for 5- and 6-dimensional Euclidean space.

In the third part, we describe some of the applications of Clifford algebras. Our intention here is to provide an introduction to a varied collection of applications; to whet the appetite, so that the reader will wish to pursue his or her interests further.

A great deal of interest in Clifford algebras goes back to 1927 and 1928. In 1927, Pauli introduced the so-called Pauli spin matrices to provide a quantum mechanical framework for particles with spin 1/2, and in 1928, Dirac introduced the Dirac operator (though not with this name, nor in terms of a Clifford algebra) to construct the Dirac equation, which describes the relativistic behaviour of an electron. In Chapter 9, we describe the use of the Pauli spin matrices to represent the angular momentum of particles with spin 1/2. We introduce the Dirac operator, and construct the Dirac equation. We also show that Maxwell's equations for electromagnetic fields can be expressed as a single equation involving the Dirac operator.

Clifford algebras have important applications in differential and harmonic analysis. A fascinating topic in two dimensions is the relationship between harmonic functions and analytic functions, using the Hilbert transform. In Chapter 10, we show how the Dirac operator, and an augmented Dirac operator, can be used to extend the idea of analyticity to higher dimensions, so that corresponding problems can be considered there; the Hilbert transform is replaced by the system of Riesz transforms. As a particular application, we show how to extend to higher dimensions the celebrated theorem of the brothers Riesz, which shows that if the harmonic Dirichlet extension of a complex measure is analytic, then the measure is absolutely continuous with respect to Lebesgue measure, and is represented by a function in the Hardy space H_1.

These results concern functions defined on half a vector space. An even more important use of Dirac operators concerns analysis on a compact Riemannian manifold; this leads to proofs of the Atiyah-Singer index theorem. A full account of this demands a detailed knowledge of Riemannian geometry, and so cannot be given at this introductory level, but we end Chapter 10 by giving a brief description of the set-up in which Dirac operators can be defined.

The spin groups provide a double cover of the special orthogonal groups. In Chapter 11, we show how this can be used to describe the irreducible representations of the special orthogonal groups of Euclidean spaces of dimensions 2, 3 and 4. The use of the double cover goes much further, but again this requires a detailed understanding of the representation theory of compact Lie groups, inappropriate for a book at this elementary level.

These remarks show that this book is only an introduction to a large subject, with many applications. In the final chapter, we make some further comments, and also make some suggestions for further reading.

I would especially like to thank the referees for their dissatisfaction with earlier drafts of this book, which led to improvements both of content and of presentation. I have worked hard to remove errors, but undoubtedly some remain. Corrections and further comments can be found on my personal web-page at www.dpmms.cam.ac.uk.

I acknowledge the use of Paul Taylor's 'diagrams' package, which I used for the commutative diagrams in the text; I found the package easy to use.

PART ONE

THE ALGEBRAIC ENVIRONMENT

1
Groups and vector spaces

The material in this chapter should be familiar to the reader, but it is worth reading through it to become familiar with the notation and terminology that is used. We shall not give details; these are given in standard textbooks, such as Mac Lane and Birkhoff [MaB], Jacobson [Jac] or Cohn [Coh].

1.1 Groups

A *group* is a non-empty set G together with a *law of composition*, a mapping $(g, h) \to gh$ from $G \times G$ to G, which satisfies:

1. $(gh)j = g(hj)$ for all g, h, j in G (associativity),
2. there exists e in G such that $eg = ge = g$ for all $g \in G$, and
3. for each $g \in G$ there exists $g^{-1} \in G$ such that $gg^{-1} = g^{-1}g = e$.

It then follows that e, the *identity element*, is unique, and that for each $g \in G$ the *inverse* g^{-1} is unique.

A group G is *abelian*, or *commutative*, if $gh = hg$ for all $g, h \in G$. If G is abelian, then the law of composition is often written as addition: $(g, h) \to g + h$. In such a case, the identity is denoted by 0, and the inverse of g by $-g$.

A non-empty subset H of a group G is a *subgroup* of G if $h_1 h_2 \in H$ whenever $h_1, h_2 \in H$, and $h^{-1} \in H$ whenever $h \in H$. H then becomes a group under the law of composition inherited from G.

If A is a subset of a group G, there is a smallest subgroup $\mathrm{Gp}(A)$ of G which contains A, the *subgroup generated by* A. If $A = \{g\}$ is a singleton then we write $\mathrm{Gp}(g)$ for $\mathrm{Gp}(A)$. Then $\mathrm{Gp}(g) = \{g^n : n \text{ an integer}\}$, where $g^0 = e$, g^n is the product of n copies of g when $n > 0$, and g^n

is the product of $|n|$ copies of g^{-1} when $n < 0$. A group G is *cyclic* if $G = \mathrm{Gp}(g)$ for some $g \in G$.

If G has finitely many elements, then the *order* $o(G)$ of G is the number of elements of G. If G has infinitely many elements, then we set $o(G) = \infty$. If $g \in G$ then the *order* $o(g)$ of g is the order of the group $\mathrm{Gp}(g)$.

A mapping $\theta : G \to H$ from a group G to a group H is a *homomorphism* if $\theta(g_1 g_2) = \theta(g_1)\theta(g_2)$, for $g_1, g_2 \in G$. It then follows that θ maps the identity in G to the identity in H, and that $\theta(g^{-1}) = (\theta(g))^{-1}$, for $g \in G$. A bijective homomorphism is called an *isomorphism*, and an isomorphism $G \to G$ is called an *automorphism* of G. The set $\mathrm{Aut}(G)$ of automorphisms of G forms a group, when composition of mappings is taken as the group law of composition.

A subgroup K of a group G is a *normal*, or *self-conjugate*, subgroup if $g^{-1}hg \in K$ for all $g \in G$ and $k \in K$. If $\theta : G \to H$ is a homomorphism, then the *kernel* $\ker(\theta)$ of θ, the set $\{g \in G : \theta g = e_H\}$ (where e_H is the identity in H) is a normal subgroup of G. Conversely, suppose that K is a normal subgroup of G. The relation $g_1 \sim g_2$ on G defined by setting $g_1 \sim g_2$ if $g_1^{-1}g_2 \in K$ is an equivalence relation on G. The equivalence classes are called the *cosets* of K in G. If C is a coset of K then C is of the form $Kg = \{kg : k \in K\}$ and $Kg = gK = \{gk : k \in K\}$. If C_1 and C_2 are cosets of K in G then so is $C_1 C_2 = \{c_1 c_2 : c_1 \in C_1, c_2 \in C_2\}$; if $C_1 = Kg_1$ and $C_2 = Kg_2$ then $C_1 C_2 = Kg_1 g_2$. With this law of composition, the set G/K of cosets becomes a group, the *quotient group*. The identity is K and $(Kg)^{-1} = Kg^{-1}$. The quotient mapping $q : G \to G/K$ defined by the equivalence relation is then a homomorphism of G onto G/K, with kernel K, and $q(g) = Kg$.

A group G is *simple* if it has no normal subgroups other than $\{e\}$ and G.

We denote the group with one element by 1, or, if we are denoting composition by addition, by 0. Suppose that $(G_0 = 1, G_1, \ldots, G_k, G_{k+1} = 1)$ is a sequence of groups, and that $\theta_j : G_j \to G_{j+1}$ is a homomorphism, for $0 \leq j \leq k$. Then the diagram

$$1 \xrightarrow{\theta_0} G_1 \xrightarrow{\theta_1} G_2 \xrightarrow{\theta_2} \ldots \xrightarrow{\theta_{k-1}} G_k \xrightarrow{\theta_k} 1$$

is an *exact sequence* if $\theta_{j-1}(G_{j-1})$ is the kernel of θ_j, for $1 \leq j \leq k$. If $k = 3$, the sequence is a *short exact sequence*. For example, if K is a normal subgroup of g and $q : G \to G/K$ is the quotient mapping, then

$$1 \longrightarrow K \overset{\subseteq}{\longrightarrow} G \overset{q}{\longrightarrow} G/K \longrightarrow 1$$

is a short exact sequence. If A is a subset of a group G then the *centralizer* $C_G(A)$ of A in G, defined as

$$C_G(A) = \{g \in G : ga = ag \text{ for all } a \in A\},$$

is a subgroup of G. The *centre* $Z(G)$, defined as

$$Z(G) = \{g \in G : gh = hg \text{ for all } h \in G\},$$

(which is $C_G(G)$), is a normal subgroup of G.

The product of two groups $G_1 \times G_2$ is a group, when composition is defined by

$$(g_1, g_2)(h_1, h_2) = (g_1 h_1, g_2 h_2)$$

We identify G_1 with the subgroup $G_1 \times \{e_2\}$ and G_2 with the subgroup $\{e_1\} \times G_2$.

Let us now list some of the groups that we shall meet later.

1. The real numbers $\mathbf{R}$ form an abelian group under addition. The set $\mathbf{Z}$ of integers is a subgroup of $\mathbf{R}$. The set $\mathbf{R}^*$ of non-zero real numbers is a group under multiplication.

2. Any two groups of order 2 are isomorphic. We shall denote the multiplicative subgroup $\{1, -1\}$ of $\mathbf{R}^*$ by D_2, and the additive group $\{0, 1\}$ by $\mathbf{Z}_2$. $\mathbf{Z}_2$ is isomorphic to the quotient group $\mathbf{Z}/2\mathbf{Z}$. Small though they are, these groups of order 2 play a fundamental role in the theory of Clifford algebras (and in many other branches of mathematics and physics).

   Suppose that we have a short exact sequence

   $$1 \longrightarrow D_2 \overset{j}{\longrightarrow} G_1 \overset{\theta}{\longrightarrow} G_2 \longrightarrow 1.$$

   Then $j(D_2)$ is a normal subgroup of G_1, from which it follows that $j(D_2)$ is contained in the centre of G_1. If $g \in G_1$ then we write $-g$ for $j(-1)g$. Then $\theta(g) = \theta(-g)$, and if $h \in G_2$ then $\theta^{-1}\{h\} = \{g, -g\}$ for some g in G. In this case, we say that G_1 is a *double cover* of G_2. Double covers play a fundamental role in the theory of spin groups; these are considered in Chapter 8.

3. A bijective mapping of a set X onto itself is called a *permutation*. The set Σ_X of permutations of X is a group under the composition of mappings. Σ_X is not abelian if X has at least three elements. We

denote the group of permutations of the set $\{1,\ldots,n\}$ by Σ_n. Σ_n has order $n!$. A *transposition* is a permutation which fixes all but 2 elements. Σ_n has a normal subgroup A_n of order $n!/2$, consisting of those permutations that can be expressed as the product of an even number of transpositions. Thus we have a short exact sequence

$$1 \longrightarrow A_n \overset{\subseteq}{\longrightarrow} \Sigma_n \overset{\epsilon}{\longrightarrow} D_2 \longrightarrow 1.$$

If $\sigma \in \Sigma_n$ then $\epsilon(\sigma)$ is the *signature* of σ; $\epsilon(\sigma) = 1$ if $\sigma \in A_n$, and $\epsilon(\sigma) = -1$ otherwise.

4. The complex numbers $\mathbf{C}$ form an abelian group under addition, and $\mathbf{R}$ can be identified as a subgroup of $\mathbf{C}$. The set $\mathbf{C}^*$ of non-zero complex numbers is a group under multiplication. The set $\mathbf{T} = \{z \in \mathbf{C} : |z| = 1\}$ is a subgroup of $\mathbf{C}^*$. There is a short exact sequence

$$0 \longrightarrow \mathbf{Z} \overset{\subseteq}{\longrightarrow} \mathbf{R} \overset{q}{\longrightarrow} \mathbf{T} \longrightarrow 1$$

where $\mathbf{Z}$ is the additive group of integers and $q(\theta) = e^{2\pi i\theta}$.

5. The subset $\mathbf{T}_n = \{e^{2\pi ij/n} : 0 \le j < n\} = \{z \in \mathbf{C} : z^n = 1\}$ of $\mathbf{T}$ is a cyclic subgroup of $\mathbf{T}$ of order n. Conversely, if $G = Gp(g)$ is a cyclic group of order n then the mapping $g^k \to e^{2\pi ik/n}$ is an isomorphism of G onto $\mathbf{T}_n$.

6. Let D denote the group of isometries of the complex plane $\mathbf{C}$ which fix the origin:

$$D = \{g : \mathbf{C} \to \mathbf{C} : g(0) = 0 \text{ and } |g(z) - g(w)| = |z - w| \text{ for } z, w \in \mathbf{C}\}.$$

D is the *full dihedral group*. Then an element of D is either a *rotation* R_θ (where $R_\theta(z) = e^{i\theta}z$) or a *reflection* S_θ (where $S_\theta(z) = e^{i\theta}\bar{z}$). The set Rot of rotations is a subgroup of D, and the mapping $R : e^{i\theta} \to R_\theta$ is an isomorphism of $\mathbf{T}$ onto Rot. In particular, $R_\pi(z) = -z$. Since

$$S_\theta^2(z) = e^{i\theta}\overline{(e^{i\theta}\bar{z})} = e^{i\theta}e^{-i\theta}z = z,$$

S_θ^2 is the identity. A similar calculation shows that $S_\theta^{-1}R_\phi S_\theta = R_{-\phi}$, so that R is a normal subgroup of D. We have an exact sequence

$$1 \longrightarrow \mathbf{T} \overset{R}{\longrightarrow} D \overset{\delta}{\longrightarrow} D_2 \longrightarrow 1$$

where $\delta(R_\theta)) = 1$ and $\delta(S_\theta) = -1$, for $\theta \in [0, 2\pi)$.

7. If $n \ge 2$, let $R_n = R(\mathbf{T}_n)$ and let $D_{2n} = R_n \cup R_n S_0$. D_{2n} is a subgroup of D, called the *dihedral group of order* $2n$. (Warning: some authors denote this group by D_n.) The group $D_4 \cong D_2 \times D_2$, and so D_4 is

abelian. If $n \geq 3$ then D_{2n} is the group of symmetries of a regular polygon with n vertices, with centre the origin; D_{2n} is a non-abelian subgroup of D of order $2n$. If $n = 2k$ is even, then $Z(D_{2n}) = \{1, r_{-1}\}$, and we have a short exact sequence

$$1 \longrightarrow D_2 \longrightarrow D_{2n} \longrightarrow D_{2k} \longrightarrow 1;$$

D_{2n} is a double cover of D_{2k}. If $n = 2k+1$ is odd, then $Z(D_{2n}) = \{1\}$.

8. In particular, the dihedral group D_8 is a non-abelian group, which is the group of symmetries of a square with centre the origin. Let us set $\alpha = r_i$, $\beta = \sigma_1$ and $\gamma = \sigma_i$. Then $D_8 = \{\pm 1, \pm\alpha, \pm\beta, \pm\gamma\}$ (where $-x$ denotes $r_{-1}x = xr_{-1}$), and

$$\begin{array}{ccc} \alpha\beta = \gamma & \beta\gamma = \alpha & \gamma\alpha = \beta \\ \beta\alpha = -\gamma & \gamma\beta = -\alpha & \alpha\gamma = -\beta \\ \alpha^2 = -1 & \beta^2 = 1 & \gamma^2 = 1. \end{array}$$

There is a short exact sequence

$$1 \longrightarrow D_2 \longrightarrow D_8 \longrightarrow D_4 \longrightarrow 1.$$

9. The *quaternionic group* $\mathcal{Q}$ is a group of order 8, with elements $\{\pm 1, \pm i, \pm j, \pm k\}$, with identity element 1, and law of composition defined by

$$\begin{array}{ccc} ij = k & jk = i & ki = j \\ ji = -k & kj = -i & ik = -j \\ i^2 = -1 & j^2 = -1 & k^2 = -1, \end{array}$$

and $(-1)x = x(-1) = -x$, $(-1)(-x) = (-x)(-1) = x$ for $x = 1, i, j, k$. Then $Z(\mathcal{Q}) = \{1, -1\}$, and there is a short exact sequence

$$1 \longrightarrow D_2 \longrightarrow \mathcal{Q} \longrightarrow D_4 \longrightarrow 1;$$

$\mathcal{Q}$ is a double cover of D_4.

The groups D_8 and $\mathcal{Q}$ are of particular importance in the study of Clifford algebras. Although they both provide double covers of D_4, they are not isomorphic. $\mathcal{Q}$ has six elements of order 4, one of order 2 and one of order 1. D_8 has two elements of order 4, five of order 2 and one of order 1.

1.2 Vector spaces

We shall be concerned with real or complex vector spaces. Let K denote either the field $\mathbf{R}$ of real numbers or the field $\mathbf{C}$ of complex numbers. A *vector space* E over K is an abelian additive group $(E, +)$, together with a mapping (*scalar multiplication*) $(\lambda, x) \to \lambda x$ of $K \times E$ into E which satisfies

- $1.x = x$,
- $(\lambda + \mu)(x) = \lambda x + \mu x$,
- $\lambda(\mu x) = (\lambda \mu)x$,
- $\lambda(x + y) = \lambda x + \lambda y$,

for $\lambda, \mu \in K$ and $x, y \in E$. The elements of E are called *vectors* and the elements of K are called *scalars*.

It then follows that $0.x = 0$ and $\lambda.0 = 0$ for $x \in E$ and $\lambda \in K$. (Note that we use the same symbol 0 for the additive identity element in E and the zero element in K.)

A non-empty subset F of a vector space E is a *linear subspace* if it is a subgroup of E and if $\lambda x \in U$ whenever $\lambda \in K$ and $x \in F$. A linear subspace is then a vector space, with the operations inherited from E. If A is a subset of E then the intersection of all the linear subspaces containing A is a linear subspace, the subspace span(A) *spanned* by A. If E is spanned by a finite set, E is *finite-dimensional*.

All the vector spaces that we shall consider will be finite-dimensional, except when it is stated otherwise.

A subset B of E is *linearly independent* if whenever $\lambda_1, \ldots, \lambda_k$ are scalars and $b_1, \ldots, b_k$ are distinct elements of B for which $\lambda_1 b_1 + \cdots + \lambda_k b_k = 0$ then $\lambda_1 = \cdots = \lambda_k = 0$.

A finite subset B of E which is linearly independent and which spans E is called a *basis* for E. We shall enumerate a basis as $(b_1, \ldots, b_d)$. If $(b_1, \ldots, b_d)$ is a basis for E then every element $x \in E$ can be written uniquely as $x = x_1 b_1 + \cdots + x_d b_d$ (where $x_1, \ldots, x_d$ are scalars). Every finite-dimensional vector space E has a basis. Indeed if A is a linearly independent subset of E contained in a subset C of E which spans E then there is a basis B for E with $A \subseteq B \subseteq C$. Any two bases have the same number of elements; this number is the *dimension* $\dim E$ of E.

As an example, let $E = K^d$, the product of d copies of K, with addition defined co-ordinatewise, and with scalar multiplication

$$\lambda(x_1, \ldots, x_d) = (\lambda x_1, \ldots, \lambda x_d).$$

Let $e_j = (0, \ldots, 0, 1, 0, \ldots, 0)$, with 1 in the jth position. Then K^d is a vector space, and $(e_1, \ldots, e_d)$ is a basis for K^d, the *standard basis*. More generally, Let $M_{m,n} = M_{m,n}(K)$ denote the set of all K-valued functions on $\{1, \ldots, m\} \times \{1, \ldots, n\}$. $M_{m,n}$ becomes a vector space over K when addition and multiplication are defined co-ordinatewise. The elements of $M_{m,n}$ are called *matrices*. We denote the matrix taking the value 1 at (i, j) and 0 elsewhere by E_{ij}. Then the set of matrices $\{E_{ij} : 1 \le i \le m, 1 \le j \le n\}$ forms a basis for $M_{m,n}$, so that $M_{m,n}$ has dimension mn.

If E_1 and E_2 are vector spaces, then the product $E_1 \times E_2$ is a vector space, with addition and scalar multiplication defined by

$$(x_1, x_2) + (y_1, y_2) = (x_1 + y_1, x_2 + y_2) \quad \text{and} \quad \lambda(x_1, x_2) = (\lambda x_1, \lambda x_2),$$

Then $\dim(E_1 \times E_2) = \dim E_1 + \dim E_2$.

A mapping $T : E \to F$, where E and F are vector spaces over the same field K, is *linear* if

$$T(x + y) = T(x) + T(y) \quad \text{and} \quad T(\lambda x) = \lambda T(x) \quad \text{for all } \lambda \in K, x, y \in E.$$

The *image* $T(E)$ is a linear subspace of F and the *null-space* $N(T) = \{x \in E : T(x) = 0\}$ is a linear subspace of E. The dimension of $T(E)$ is the *rank* of T and the dimension of $N(T)$ is the *nullity* $n(T)$ of T. We have the fundamental *rank-nullity formula*

$$\text{rank}(T) + n(T) = \dim E.$$

A bijective linear mapping $J : E \to F$ is called an *isomorphism*. A linear mapping $J : E \to F$ is an isomorphism if and only if $N(J) = \{0\}$ and $J(E) = F$. If J is an isomorphism, then $\dim E = \dim F$. For example, if $(f_1, \ldots, f_d)$ is a basis for F then the linear mapping $J : K^d \to F$ defined by $J(\lambda_1, \ldots, \lambda_d) = \lambda_1 f_1 + \cdots + \lambda_d f_d$ is an isomorphism of $\mathbf{K}^d$ onto F.

We shall occasionally need to consider the topology on a vector space F. K^d is a complete metric space, with the usual metric $d(x, y) = (\sum_{j=1}^d |x_j - y_j|^2)^{1/2}$ given by the Euclidean norm $\|x\| = (\sum_{j=1}^d |x_j|^2)^{1/2}$. We can then define a norm $\|.\|$ on F by setting $\|J(x)\| = \|x\|$. This depends on the choice of basis, but any two such norms are equivalent, and define the same topology.

If F_1 and F_2 are subspaces of E for which the linear mapping $(x_1, x_2) \to x_1 + x_2 : F_1 \times F_2 \to E$ is an isomorphism then E is the *direct sum* $F_1 \oplus F_2$ of F_1 and F_2. This happens if and only if $F_1 \cap F_2 = \{0\}$ and $E = \text{span}(F_1 \cup F_2)$, and if and only if every element x of E can be written uniquely as $x = x_1 + x_2$, with $x_1 \in F_1$ and $x_2 \in F_2$.

Suppose that $(e_1, \ldots, e_d)$ is a basis for E and that $y_1, \ldots, y_d$ are elements of a vector space F. If $x = \lambda_1 e_1 + \cdots + \lambda_d e_d \in E$, let $T(x) = \lambda_1 y_1 + \cdots + \lambda_d y_d$. Then T is a linear mapping of E into F, and is the unique linear mapping from E to F for which $T(e_j) = y_j$ for $1 \le j \le d$. The process of constructing T in this way is called *extension by linearity*.

The set $L(E, F)$ of linear mappings from E to F is a vector space, when we define

$$(S + T)(x) = S(x) + T(x) \quad \text{and} \quad (\lambda S)(x) = \lambda(S(x))$$

for $S, T \in L(E, F)$, $x \in E$, $\lambda \in K$. $\dim L(E, F) = \dim E . \dim F$. We write $L(E)$ for $L(E, E)$; elements of $L(E)$ are called *endomorphisms* of E.

If $T \in L(E, F)$ and $S \in L(F, G)$ then the composition $ST = S \circ T$ is in $L(E, G)$.

Suppose that $(e_1, \ldots, e_d)$ is a basis for E, $(f_1, \ldots, f_c)$ a basis for F and that $T \in L(E, F)$. Let $T(e_j) = \sum_{i=1}^{c} t_{ij} f_i$. If $x = \sum_{j=1}^{d} x_j e_j$ then $T(x) = \sum_{i=1}^{c} (\sum_{j=1}^{d} t_{ij} x_j) f_i$. The mapping $T \to (t_{ij})$ is then an isomorphism of $L(E, F)$ onto $M_{c,d}$, so that $\dim L(E, F) = cd = \dim E . \dim F$. We say that T is represented by the $c \times d$ matrix (t_{ij}). If $(g_1, \ldots, g_b)$ is a basis for G, and $S \in L(F, G)$ is *represented by* the matrix (s_{hi}) then the product $R = ST \in L(E, G)$ is represented by the matrix (r_{hj}), where $r_{hj} = \sum_{i=1}^{c} s_{hi} t_{ij}$. This expression defines matrix multiplication.

1.3 Duality of vector spaces

K is a one-dimensional vector space over K (but $\mathbf{C}$ is a two-dimensional vector space over $\mathbf{R}$, with basis $\{1, i\}$). The space $L(E, K)$ is called the *dual*, or *dual space*, of E, and is denoted by E'. Elements of E' are called *linear functionals* on E. Suppose that $(e_1, \ldots, e_d)$ is a basis for E. If $x = \sum_{i=1}^{d} x_i e_i$, let $\phi_i(x) = x_i$, for $1 \le i \le d$. Then $\phi_i \in E'$ and $(\phi_1, \ldots, \phi_d)$ is a basis for E', the *dual basis*, or *basis dual to* $(e_1, \ldots, e_d)$. Thus $\dim E = \dim E'$. If $x \in E$ and $\phi \in E'$, let $j(x)(\phi) = \phi(x)$. Then

$j : E \rightarrow E''$ is an isomorphism of E onto E'', the dual of E'. E'' is called the *bidual* of E.

Suppose that $T \in L(E, F)$. If $\psi \in F'$ and $x \in E$, let $(T'(\psi))(x) = \psi(T(x))$. Then $T(\psi) \in E'$, and T' is a linear mapping of F' into E'; it is the *transposed* mapping of T.

If A is a subset of E, then the *annihilator* $A^\perp$ in E' of A is the set

$$A^\perp = \{\phi \in E' : \phi(a) = 0 \text{ for all } a \in A\}.$$

It is a linear subspace of E'. Similarly, if B is a subset of E', then the *annihilator* $B^\perp$ in E of B is the set

$$B^\perp = \{x \in E : \phi(x) = 0 \text{ for all } \phi \in B\}.$$

It then follows that $A^{\perp\perp} = \text{span}(A)$ and $B^{\perp\perp} = \text{span}(B)$. If F is a linear subspace of E then $\dim F + \dim F^\perp = \dim E$.

If $T \in L(E, F)$ then $(T(E))^\perp = N(T')$ and $(N(T))^\perp = T'(F')$.

(Several of these results depend on the fact that E and F are finite-dimensional.)

2

Algebras, representations and modules

Clifford algebras are finite-dimensional algebras. Here we consider the properties of finite-dimensional algebras. We also consider how they can be represented as algebras of endomorphisms of a vector space, or equivalently as algebras of matrices. An alternative way of thinking about this is to consider modules over an algebra; this is important in the theory of Clifford algebras, where such modules appear as spaces of spinors.

2.1 Algebras

Again, let K denote either the field $\mathbf{R}$ of real numbers or the field $\mathbf{C}$ of complex numbers. A finite-dimensional (associative) *algebra* A over K is a finite-dimensional vector space over K equipped with a law of composition: that is, a mapping (*multiplication*) $(a, b) \to ab$ from $A \times A$ into A which satisfies

- $(ab)c = a(bc)$ (associativity),
- $a(b + c) = ab + ac$,
- $(a + b)c = ac + bc$,
- $\lambda(ab) = (\lambda a)b = a(\lambda b)$,

for $\lambda \in K$ and $a, b, c \in A$. (As usual, multiplication is carried out before addition).

An algebra A is *unital* if there exists $1 \in A$, the *identity element*, such that $1a = a1 = a$ for all $a \in A$. We shall principally be concerned with unital algebras. An algebra A is *commutative* if $ab = ba$ for all $a, b \in A$.

A mapping ϕ from an algebra A over K to an algebra B over K is an *algebra homomorphism* if it is linear, and if $\phi(ab) = \phi(a)\phi(b)$ for $a, b \in A$. If A and B are unital, ϕ is a *unital homomorphism* if, in addition, $\phi(1_A) = 1_B$, where 1_A is the identity element of A and 1_B is the identity element of B. A bijective homomorphism is called an *algebra isomorphism*. An algebra homomorphism of an algebra A into itself is called an *endomorphism*.

Let us give some examples.

- If E is a vector space over K then the vector space $L(E)$ of endomorphisms of E becomes a unital algebra over K when multiplication is defined to be the composition of mappings. The identity mapping I is the algebra identity. $L(E)$ is commutative if and only if $\dim V = 1$.

- If $(e_1, \ldots, e_d)$ is a basis for E, then an element T of $L(E)$ can be represented by a matrix (t_{ij}), and the mapping $T \to (t_{ij})$ is an algebra isomorphism of $L(E)$ onto the algebra $M_d(K)$ of $d \times d$ matrices, where composition is defined as matrix multiplication.

- More generally, if A is an algebra, then the set $M_d(A)$ of $d \times d$ matrices with entries in A becomes an algebra when addition and scalar multiplication are defined term by term, and composition is defined by matrix multiplication: if $a = (a_{ij})$ and $b = (b_{ij})$ then $(ab)_{ij} = \sum_{k=1}^d a_{ik}b_{kj}$. The algebra $M_d(A)$ is unital if A is.

- Let K^S denote the set of mappings from a finite set S into K. K^S becomes a commutative unital algebra when we define addition, scalar multiplication and multiplication pointwise:

$$(f + g)(s) = f(s) + g(s); \quad (\lambda f)(s) = \lambda(f(s)); \quad (fg)(s) = f(s)g(s).$$

The identity element is the function 1 which takes the constant value 1: $1(s) = 1$ for all $s \in S$. If $s \in S$, let $\delta_s(s) = 1$ and $\delta_s(t) = 0$ for $t \neq s$. Then the set of mappings $\{\delta_s : s \in S\}$ is a basis for K^S, so that $\dim K = |S|$, where $|S|$ is the number of elements in S.

- Suppose that G is a finite group, with identity element e. Then K^G is a finite-dimensional vector space, with basis $\{\delta_g : g \in G\}$. We define multiplication on K^G; if $a = \sum_{g \in G} a_g \delta_g$ and $b = \sum_{g \in G} b_g \delta_g$ then we set $ab = \sum_{g \in G} c_g \delta_g$, where

$$c_g = \sum_{hj=g} a_h b_j = \sum_{h \in G} a_h b_{h^{-1}g} = \sum_{j \in G} a_{gj^{-1}} b_j.$$

It is straightforward to verify that the axioms are satisfied: K^G is the *group algebra*. Multiplication is defined in such a way that $\delta_g \delta_h = \delta_{gh}$. In particular, K^G is a unital algebra, with identity element δ_e. K^G is commutative if and only if G is commutative. Note that this multiplication is quite different from the one described in the previous example.

- If A is an algebra, then the *opposite algebra* A^{opp} is the algebra obtained by keeping addition and scalar multiplication the same, but by defining a new law of composition $*$ by reversing the original law, so that $a * b = ba$.

A linear subspace B of an algebra A is a *subalgebra* of A if $b_1 b_2 \in B$ whenever $b_1, b_2 \in B$. If A is unital, then a subalgebra B is a unital subalgebra if the identity element of A belongs to B. For example, if A is a unital algebra then the set $\mathrm{End}(A)$ of unital endomorphisms of A is a unital subalgebra of $L(A)$.

The *centralizer* $C_A(B)$ of a subset B of an algebra A is

$$C_A(B) = \{a \in A : ab = ba \text{ for all } b \in B\},$$

and the *centre* $Z(A)$ of A is the centralizer $C_A(A)$:

$$Z(A)) = \{a \in A : ab = ba \text{ for all } b \in A\}.$$

$Z(A)$ is a commutative subalgebra of A, and is a unital subalgebra if A is a unital algebra. A unital algebra is *central* if $Z(A)$ is the one-dimensional subspace $\mathrm{span}(1)$.

An element p of an algebra A is an *idempotent* if $p^2 = p$. If p is an idempotent of a unital algebra A then $1 - p$ is also idempotent, and $A = pA \oplus (1-p)A$, the direct sum of linear subspaces of A. If in addition $p \in Z(A)$ then pA and $(1-p)A$ are subalgebras of A, and are unital algebras, with identity elements p and $1 - p$ respectively, but are not unital subalgebras of A (unless $p = 1$ or $p = 0$). The mapping $a \to pa$ is then a unital algebra homomorphism of A onto pA. An idempotent in $L(E)$ or $M_d(K)$ is called a *projection*.

An element j of a unital algebra A is an *involution* if $j^2 = 1$. An element j of A is an involution if and only if $(1+j)/2$ is an idempotent.

Similarly, an endomorphism θ of an algebra A is an involution if $\theta^2 = I$. If θ is an involution of a unital algebra A, and $p = (I+\theta)/2$ then we can write $A = A^+ \oplus A^-$, where $A^+ = p(A)$ and $A^- = (I - p)(A)$. Then $p(A)$ is a subalgebra of A,

$$A^+ = \{a \in A : \theta(a) = a\} \text{ and } A^- = \{a \in A : \theta(a) = -a\},$$

and

$$A^+.A^+ \subseteq A^+, \quad A^-.A^- \subseteq A^+, \quad A^-.A^+ \subseteq A^-, \quad A^-.A^+ \subseteq A^-. \qquad (*)$$

Conversely, if $A = A^+ \oplus A^-$ is a direct sum decomposition for which $(*)$ holds, then the mapping which sends $a^+ + a^-$ to $a^+ - a^-$ is an involution of A. An algebra for which this holds is called a $\mathbf{Z}_2$-*graded algebra* or a *super-algebra*. The elements of A^+ are called *even* elements, and the non-zero elements of A^- are called *odd* elements. Any element a of A can be decomposed as $a = a^+ + a^-$, the sum of the even and odd parts of a. If $a \in A^+ \cup A^-$, we say that a is *homogeneous*.

As an example, the real algebra $\mathbf{C}$ is a super-algebra, when we define the involution j as $j(x + iy) = x - iy$. As we shall see, Clifford algebras are super-algebras.

If a is an element of a unital algebra A then b is a *left inverse* of a if $ba = 1$. A *right inverse* is defined similarly. a is *invertible* if it has a left inverse and a right inverse; they are then unique, and equal. This unique element is called the *inverse* of a, and is denoted by a^{-1}. The set of invertible elements is denoted by $G(A)$; it is a group, under composition.

When $A = L(E)$, then $G(A)$ is denoted by $GL(E)$; it is called the *general linear group* of E. Similarly we denote $G(M_d(K))$ by $GL_d(K)$.

Proposition 2.1.1 *Suppose that $T \in L(E)$. The following are equivalent.*

(i) T is invertible.

(ii) T has a left inverse.

(iii) T has a right inverse.

Proof Clearly (i) implies (ii) and (iii). If T has a left inverse, then T is injective. It therefore follows from the rank-nullity formula that

$$\operatorname{rank}(T) = \operatorname{rank}(T) + n(T) = \dim E,$$

so that T is surjective. Thus T is bijective. The inverse mapping T^{-1} is linear, and is the inverse of T in $L(E)$. Thus (ii) implies (i). The proof that (iii) implies (i) is similar; if T has a right inverse, then T is surjective, and it follows from the rank-nullity formula that T is injective. □

There is a unique mapping, the *determinant*, det $: M_d(K) \to K$ satisfying

$$\det(ST) = \det S . \det T, \quad \det I = 1, \quad \det(\lambda T) = \lambda^d \det T,$$

for $S, T \in M_d(K)$ and $\lambda \in K$. Then $T \in GL_d(K)$ if and only if $\det T \neq 0$.

A subalgebra J of an algebra A is a *left ideal* if $aj \in J$ whenever $a \in A$ and $j \in J$; a *right ideal* is defined similarly. A subalgebra which is both a left ideal and a right ideal is called a *two-sided ideal*, or more simply an *ideal*. If $\phi : A \to B$ is an algebra homomorphism, then the null-space $N(\phi) = \{a \in A : \phi(a) = 0\}$ is an ideal in A. An ideal J in A is *proper* if J is a proper subset of A. An ideal J in a unital algebra is proper if and only if $1_A \notin J$. An algebra A is *simple* if the only proper ideal in A is the *trivial ideal* $\{0\}$.

Proposition 2.1.2　$M_d(K)$ *is a simple algebra.*

Proof　Suppose that J is an ideal in $M_d(K)$ which is not equal to $\{0\}$. There exists a non-zero element $t = (t_{ij})$ in J, and so there exist indices i' and j' such that $t_{i'j'} \neq 0$. Since J is an ideal, the matrix $E_{ii} = (1/t_{i'j'})E_{ii'}tE_{j'i} \in J$, for $i \leq i \leq d$, and so $I = \sum_{i=1}^{d} E_{ii} \in J$. Thus J is not a proper ideal in $M_d(K)$.　　　　　□

The unital algebra $M_d(K)$ is a concrete example of an algebra, as to a lesser extent is $L(E)$. A unital homomorphism π from a unital algebra into $M_d(K)$ or $L(E)$ is called a *representation* of A. A linear subspace F of E is π-*invariant* if $\pi(a)(F) \subseteq F$ for each $a \in A$. The representation is *irreducible* if $\{0\}$ and F are the only π-invariant subspaces of E. If π is one-one, the representation is said to be *faithful*. A faithful representation of A is therefore a unital isomorphism of A onto a subalgebra of $M_d(K)$; the elements of A are represented as matrices. As an important example, a faithful representation, the *left regular representation* $l : A \to L(A)$ of a unital algebra A is given by setting $l(a)(b) = ab$. It is straightforward to verify that l is a representation. Since $l_a(1) = a$, l is faithful.

Suppose that A is a finite-dimensional real unital algebra, and that $|.|$ is a norm on A. We can define another norm on A by setting $\|a\| = \sup\{|ab| : |b| \leq 1\}$; then $\|I\| = 1$ and $\|ab\| \leq \|a\| . \|b\|$. It then follows that if $a \in A$ then the sum $\sum_{j=0}^{\infty} a^j/j!$ converges. We denote the sum by e^a. The mapping $a \to e^a$ from A to A is called the *exponential function*. This will be useful in later calculations.

Proposition 2.1.3　*Suppose that A is a finite-dimensional real unital algebra and that $a, b \in A$.*
　(i) *The exponential function is continuous.*
　(ii) *If $ab = ba$ then $e^{a+b} = e^a e^b$.*
　(iii) *e^a is invertible, with inverse e^{-a}.*
　(iv) *The mapping $t \to e^t$ is a continuous homomorphism of the group $(\mathbf{R}, +)$ into $G(A)$.*
　(v) *If $ab = -ba$ then $e^a b = b e^{-a}$.*

(vi) If $a^2 = -1$ then $e^{ta} = \cos t.I + \sin t.a$, the mapping $e^{it} \to e^{at}$ is a homeomorphism of $\mathbf{T}$ onto a compact subgroup of $G(A)$, and the mapping $t \to e^{ta}$ from $[0, \pi/2]$ into $G(A)$ is a continuous path from I to a.

(vii) Suppose that $a \neq I$ and that $a^2 = I$. Then $e^{ta} = \cosh t.I + \sinh t.a$, and the mapping $t \to e^{at}$ is a homeomorphism of $\mathbf{R}$ onto an unbounded subgroup of $G(A)$.

Proof Let us give a sketch of the proof.

(i) Since

$$(a + h)^j - a^j = ((a + h)^{j-1} - a^{j-1})a - (a + h)^{j-1}h,$$

an inductive argument shows that

$$\left\| (a + h)^j - a^j \right\| \leq j \, \|h\| \, (\|a\| + \|h\|)^{j-1},$$

so that

$$\left\| \sum_{j=0}^{n} \frac{(a+h)^j}{j!} - \sum_{j=0}^{n} \frac{a^j}{j!} \right\| \leq \|h\| \sum_{j=0}^{n-1} \frac{(\|a\| + \|h\|)^j}{j!} \leq \|h\| \, e^{\|a\| + \|h\|}.$$

Letting $n \to \infty$,

$$\left\| e^{a+h} - e^a \right\| \leq \|h\| \, e^{\|a\| + \|h\|},$$

which establishes continuity.

(ii) The fact that $\sum_{j=0}^{\infty} \left\| a^j/j! \right\| \leq e^{\|a\|}$ and that a similar inequality holds for b justifies the change in the order of summation in the following equalities:

$$e^{a+b} = \sum_{j=0}^{\infty} \frac{(a+b)^j}{j!} = \sum_{j=0}^{\infty} \left(\sum_{k+l=j} \frac{a^k b^l}{k! l!} \right)$$

$$= \left(\sum_{k=0}^{\infty} \frac{a^k}{k!} \right) \left(\sum_{l=0}^{\infty} \frac{b^l}{l!} \right) = e^a e^b.$$

The remaining results follow easily from these results. For example if $ab = -ba$ then $a^j b = b(-a)^j$, so that

$$e^a b = \sum_{j=0}^{\infty} \frac{a^j b}{j!} = \sum_{j=0}^{\infty} \frac{b(-a)^j}{j!} = be^{-a}.$$

$\square$

2.2 Group representations

We can also consider the representation of groups. A homomorphism π from a group G into $GL_d(K)$ is called a *representation* of G, and π is said to be *faithful* if it is injective. A faithful representation of G is therefore an isomorphism of G onto a subgroup of $GL_d(K)$; the elements of G are represented as invertible matrices.

For example, the mapping $\pi : D \to M_2(\mathbf{R})$ defined by

$$\pi(R_\theta) = \begin{bmatrix} \cos\theta & -\sin\theta \\ \sin\theta & \cos\theta \end{bmatrix} \text{ and } \pi(S_\theta) = \begin{bmatrix} \cos\theta & \sin\theta \\ \sin\theta & -\cos\theta \end{bmatrix}$$

is a faithful representation of the full dihedral group as a group of reflections and rotations of $\mathbf{R}^2$. If we restrict this to D_8 we have a faithful representation of D_8:

$$\pi(1) = I, \qquad \pi(R_{\pi/2}) = J, \qquad \pi(S_0) = U, \qquad \pi(S_{\pi/2}) = Q,$$
$$\pi(R_\pi) = -I, \quad \pi(R_{3\pi/2}) = -J, \quad \pi(S_\pi) = -U, \quad \pi(S_{3\pi/2}) = -Q,$$

where the matrices and their actions are described in the table on the next page.

The matrices Q and U play a symmetric role. Let

$$W = \pi(R_{\pi/4}) = \frac{1}{\sqrt{2}} \begin{bmatrix} 1 & -1 \\ 1 & 1 \end{bmatrix}.$$

Then the mapping $g \to W^{-1}gW$ is an automorphism of $\pi(D_8)$ and

$$W^{-1}QW = U, \quad W^{-1}UW = -Q, \quad W^{-1}JW = J.$$

Thus, with appropriate changes of sign, the matrices Q and U can be interchanged.

Name	Matrix	Action
I	$\begin{bmatrix} 1 & 0 \\ 0 & 1 \end{bmatrix}$	Identity
$-I$	$\begin{bmatrix} -1 & 0 \\ 0 & -1 \end{bmatrix}$	Rotation by π
J	$\begin{bmatrix} 0 & -1 \\ 1 & 0 \end{bmatrix}$	Rotation by $\pi/2$
$-J$	$\begin{bmatrix} 0 & 1 \\ -1 & 0 \end{bmatrix}$	Rotation by $-\pi/2$
U	$\begin{bmatrix} 1 & 0 \\ 0 & -1 \end{bmatrix}$	Reflection in the direction $(0,1)$ with mirror $y = 0$
$-U$	$\begin{bmatrix} -1 & 0 \\ 0 & 1 \end{bmatrix}$	Reflection in the direction $(1,0)$ with mirror $x = 0$
Q	$\begin{bmatrix} 0 & 1 \\ 1 & 0 \end{bmatrix}$	Reflection in the direction $(1,-1)$ with mirror $y = x$
$-Q$	$\begin{bmatrix} 0 & -1 \\ -1 & 0 \end{bmatrix}$	Reflection in the direction $(1,1)$ with mirror $y = -x$

[Reflections are described more fully in Section 4.8.]

These matrices will play a fundamental role in all our constructions. They satisfy the relations

$$Q^2 = U^2 = J^4 = I, \quad QU = -UQ = J,$$

$$QJ = -JQ = U, \quad JU = -UJ = Q.$$

An equally important example is the faithful representation of the quaternionic group $\mathcal{Q}$ in $M_2(\mathbf{C})$, which is defined using the *associate Pauli matrices* τ_0, τ_1, τ_2, τ_3. These and the *Pauli spin matrices* σ_0, σ_1, σ_2, σ_3 are described in the following table.

Name	Matrix	Name	Matrix
$\sigma_0 = I$	$\begin{bmatrix} 1 & 0 \\ 0 & 1 \end{bmatrix}$	$\tau_0 = I$	$\begin{bmatrix} 1 & 0 \\ 0 & 1 \end{bmatrix}$
$\sigma_1 = \sigma_x = Q$	$\begin{bmatrix} 0 & 1 \\ 1 & 0 \end{bmatrix}$	$\tau_1 = \tau_x = iQ$	$\begin{bmatrix} 0 & i \\ i & 0 \end{bmatrix}$
$\sigma_2 = \sigma_y = iJ$	$\begin{bmatrix} 0 & -i \\ i & 0 \end{bmatrix}$	$\tau_2 = \tau_y = -J$	$\begin{bmatrix} 0 & 1 \\ -1 & 0 \end{bmatrix}$
$\sigma_3 = \sigma_z = U$	$\begin{bmatrix} 1 & 0 \\ 0 & -1 \end{bmatrix}$	$\tau_3 = \tau_z = iU$	$\begin{bmatrix} i & 0 \\ 0 & -i \end{bmatrix}$

[Notation and conventions vary; here we follow Dirac [Dir].]

The representation is defined by setting

$$\pi(1) = \tau_0, \qquad \pi(\boldsymbol{i}) = \tau_1, \qquad \pi(\boldsymbol{j}) = \tau_2, \qquad \pi(\boldsymbol{k}) = \tau_3,$$
$$\pi(-1) = -\tau_0, \quad \pi(-\boldsymbol{i}) = -\tau_1, \quad \pi(-\boldsymbol{j}) = -\tau_2 \quad \pi(-\boldsymbol{k}) = -\tau_3.$$

Exercise

1. Show that the Pauli spin matrices generate a group G of order 16. What is its centre? What is the quotient group $G/\{I, -I\}$?

2.3 The quaternions

Both $\mathbf{R}$ and $\mathbf{C}$ can be considered as finite-dimensional real algebras: $\mathbf{R}$ has real dimension 1 and $\mathbf{C}$ has real dimension 2. Each is a field: it is commutative, and every non-zero element has a multiplicative inverse. We can however drop the commutativity requirement: a *division algebra* is an algebra in which every non-zero element has a multiplicative inverse. The algebra $\mathbf{H}$ of *quaternions* provides an example of a non-commutative finite-dimensional real division algebra.

This algebra was famously invented by Hamilton in 1843 after ten years of struggle. It is denoted by $\mathbf{H}$, in his honour. There are many ways of constructing this algebra; let us use the associate Pauli matrices τ_0, τ_1, τ_2 and τ_3. We can consider $M_2(\mathbf{C})$ as an eight-dimensional real algebra. The associate Pauli matrices form a linearly independent subset of $M_2(\mathbf{C})$, and so their real linear span is a four-dimensional subspace H. If $h = a\tau_0 + b\tau_1 + c\tau_2 + d\tau_3 \in H$ then

$$h = \begin{bmatrix} a + id & c + ib \\ -c + ib & a - id \end{bmatrix}.$$

Consequently

$$H = \left\{ \begin{bmatrix} z & w \\ -\bar{w} & \bar{z} \end{bmatrix} : z, w \in \mathbf{C} \right\}.$$

Since composition of the associate Pauli matrices generates the group $\pi(\mathcal{Q})$, which is contained in H, it follows that H is a four-dimensional unital subalgebra of the eight-dimensional real algebra $M_2(\mathbf{C})$.

We now define the algebra $\mathbf{H}$ of quaternions to be any real algebra which is isomorphic as an algebra to H. Suppose that $\phi : H \to \mathbf{H}$ is an isomorphism. We set

$$1 = \phi(\tau_0), \ \boldsymbol{i} = \phi(\tau_1), \ \boldsymbol{j} = \phi(\tau_2), \ \text{and} \ \boldsymbol{k} = \phi(\tau_3).$$

Then of course it follows that these elements of $\mathbf{H}$ satisfy the relations

$$\begin{array}{ccc} \boldsymbol{ij} = \boldsymbol{k} & \boldsymbol{jk} = \boldsymbol{i} & \boldsymbol{ki} = \boldsymbol{j} \\ \boldsymbol{ji} = -\boldsymbol{k} & \boldsymbol{kj} = -\boldsymbol{i} & \boldsymbol{ik} = -\boldsymbol{j} \\ \boldsymbol{i}^2 = -1 & \boldsymbol{j}^2 = -1 & \boldsymbol{k}^2 = -1, \end{array}$$

and $\{\pm 1, \pm\boldsymbol{i}, \pm\boldsymbol{j}, \pm\boldsymbol{k}\}$ is a multiplicative group isomorphic to the quaternionic group $\mathcal{Q}$.

Elements of span(1) are called *real quaternions* and elements of span($\boldsymbol{i}, \boldsymbol{j}, \boldsymbol{k}$) are called *pure quaternions*; the space of pure quaternions is denoted by $Pu(\mathbf{H})$. We can characterize these in the following way.

Proposition 2.3.1 *(i) A quaternion x is real if and only if it is in the centre $Z(\mathbf{H})$ of $\mathbf{H}$.*

(ii) A quaternion x is pure if and only if x^2 is real and non-positive.

Proof (i) Certainly the real quaternions are in the centre of $\mathbf{H}$. Conversely, if $x = a1 + bi + cj + dk \in Z(\mathbf{H})$ then

$$ai - b1 + ck - dj = ix = xi = ai - b1 - ck + dj,$$

so that $c = d = 0$, and a similar argument, considering jx, shows that $b = 0$; thus x is real.

(ii) If $x = bi + cj + dk$ is pure, then $x^2 = -b^2 - c^2 - d^2$, so that the condition is necessary. On the other hand, if $x = a1 + y$, where y is pure, then $x^2 = (a^2 - y^2) + 2ay$, so that if x^2 is real and non-positive then $ay = 0$. If $a = 0$ then x is pure. If $y = 0$ then $x^2 = a^2$, so that $a = 0$, and $x = 0$. $\square$

We now define *conjugation*. If $x = a1 + y = a1 + bi + cj + dk$, we set $\bar{x} = a1 - y = a1 - bi - cj - dk$. Conjugation is a linear involution ($\bar{\bar{x}} = x$) and $\overline{x_1 x_2} = \bar{x}_2 . \bar{x}_1$. Thus the mapping $x \to \bar{x}$ is an algebra isomorphism of $\mathbf{H}$ onto $\mathbf{H}^{opp}$.

We set $\Delta(x) = x\bar{x} = \bar{x}x = a^2 - y^2 = a^2 + b^2 + c^2 + d^2$; Δ is the *quadratic norm* on $\mathbf{H}$. We set $\|x\| = (\Delta(x))^{1/2}$; clearly $\|.\|$ is a norm on $\mathbf{H}$, and the mapping $(a, b, c, d) \to a1 + bi + cj + dk$ is a linear isometry of $\mathbf{R}^4$, with its Euclidean norm, onto $\mathbf{H}$. We denote the corresponding inner product on $\mathbf{H}$ by $\langle ., . \rangle$, and say that x and y are *orthogonal* if $\langle x, y \rangle = 0$. (Euclidean norms, inner products and orthogonality are described in Chapter 4.)

Proposition 2.3.2 *If $x, y \in Pu(\mathbf{H})$ then x is orthogonal to y if and only if $xy = -yx$.*

Proof For x is orthogonal to y if and only if $\Delta(x + y) = \Delta(x) + \Delta(y)$, if and only if $(x + y)^2 = x^2 + y^2$, if and only if $xy + yx = 0$. $\square$

Theorem 2.3.1 $\mathbf{H}$ *is a division algebra.*

Proof If $x \neq 0$ then $\Delta(x) > 0$, and

$$x(\bar{x}/\Delta(x)) = (\bar{x}/\Delta(x))x = 1,$$

so that x is invertible, with inverse $\bar{x}/\Delta(x)$. $\square$

Let $\mathbf{H}^* = \mathbf{H} \setminus \{0\}$. $\mathbf{H}^*$ is a group, under multiplication.

Proposition 2.3.3 *The mapping $\Delta : \mathbf{H}^* \to \mathbf{R}^*$ is a homomorphism of $\mathbf{H}^*$ onto the multiplicative group $\mathbf{R}^* = \{\lambda \in \mathbf{R} : \lambda \neq 0\}$.*

Proof For $\Delta(xy) = xy\overline{(xy)} = xy\bar{y}\bar{x} = \Delta(y)x\bar{x} = \Delta(x)\Delta(y)$. $\square$

The kernel of this homomorphism is $\mathbf{H}_1 = \{x \in \mathbf{H} : \Delta x = 1\}$, the unit sphere of $\mathbf{H}$. The set $\mathbf{H}_1$ is bounded and closed in the normed space $\mathbf{H}$, and is therefore compact. The group operations are continuous, and so $\mathbf{H}_1$ is an example of a compact topological group.

Hamilton invented quaternions, because he wished to extend the multiplication of complex numbers to higher dimensions. Can we go further?

Theorem 2.3.2 (Frobenius' theorem) *The algebras* $\mathbf{R}$, $\mathbf{C}$ *and* $\mathbf{H}$ *are the only finite-dimensional real division algebras.*

Proof We shall defer proof of this until Section 6.1; the proof uses ideas and results from the theory of Clifford algebras. $\qquad\square$

Why else is the algebra $\mathbf{H}$ of quaternions of interest? We shall see in Section 4.9 that they can be used to describe the groups of rotations in three and four dimensions. We shall see that $\mathbf{H}$ is an example of a Clifford algebra, that it plays an essential role in the structure of Clifford algebras, and that Clifford algebras, in their turn, describe groups of rotations in higher dimensions.

When A is a real unital algebra, we shall use the term 'representation' in a more general way than in Section 2.1. The complex algebra $M_d(\mathbf{C})$ can be considered as a real algebra of dimension $2d^2$, and the algebra $M_d(\mathbf{H})$ can be considered as a real algebra of dimension $4d^2$. We shall also call a unital homomorphism π of A into $M_d(\mathbf{C})$ or $M_d(\mathbf{H})$ a *representation* of the algebra A.

For example, the mapping $\pi : \mathbf{C} \to M_2(\mathbf{R})$ defined by

$$\pi(x + iy) = xI + yJ = \begin{bmatrix} x & y \\ -y & x \end{bmatrix}$$

is a faithful representation of the real algebra $\mathbf{C}$ in $M_2(\mathbf{R})$, and the mapping $\rho : \mathbf{H} \to M_2(\mathbf{C})$ defined by

$$\rho(a1 + b\boldsymbol{i} + c\boldsymbol{j} + d\boldsymbol{k}) = aI + ibQ - cJ + idU = \begin{bmatrix} a + id & c + ib \\ -c + ib & a - id \end{bmatrix}$$

is a faithful representation of the real algebra $\mathbf{H}$ in the real algebra $M_2(\mathbf{C})$.

The reason for considering such representations is that, as we shall see, every Clifford algebra is either isomorphic to $M_d(\mathbf{D})$, where $\mathbf{D} = \mathbf{R}, \mathbf{C}$ or $\mathbf{H}$, or is isomorphic to a direct sum $M_d(\mathbf{D}) \oplus M_d(\mathbf{D})$. In the rest of this chapter, we shall show that this follows from the fact that Clifford algebras are either simple or the direct sum of two simple subalgebras.

In fact, we establish these isomorphisms directly, and the rest of this chapter can be omitted, at least on a first reading.

2.4 Representations and modules

We introduced the notion of the representation of an algebra in Section 2.1. We can look at representations in a rather different way. As an example, if $\theta : A \to L(E)$ is a representation of A, then we can think of the endomorphisms $\theta(a) : E \to E$ as extensions of the scalars acting on E. This leads to the following definition. Suppose that A is a unital algebra over the real field $\mathbf{R}$. A *left A-module* M is a real vector space M together with a multiplication mapping $(a, m) \to am$ from $A \times M$ to M which is *bilinear*:

$$(\lambda_1 a_1 + \lambda_2 a_2)m = \lambda_1(a_1 m) + \lambda_2(a_2 m)$$

$$a(\mu_1 m_1 + \mu_2 m_2) = \mu_1(am_1) + \mu_2(am_2),$$

and satisfies

$$(ab)m = a(bm) \text{ and } 1_A m = m,$$

for all $a, a_1, a_2, b \in A, \lambda_1, \lambda_2, \mu_1, \mu_2 \in \mathbf{R}$.

A *right A-module* is defined similarly, with bilinear multiplication $M \times A \to A$ satisfying $m(ab) = (ma)b$. Note however that if M is a right A-module then M can be considered as a left A^{opp}-module, defining $am = ma$. For this reason, we shall only consider left A-modules.

We shall develop some of the theory of modules: enough for our needs. The general theory applies to a much more general situation.

If M is a left A-module, let $\theta(a)(m) = am$; then θ is a representation of A in $L(M)$. Conversely, if θ is a representation of A in $L(E)$ then E becomes a left A-module when we define $ax = \theta(a)(x)$. Thus there is a one-one correspondence between left A-modules and representations of A.

In particular, the left regular representation l of A corresponds to considering A as a left A-module, using the algebra multiplication to define the module multiplication.

We can form the *direct sum* of left A-modules: if M_1 and M_2 are left A-modules, then the vector space sum $M_1 \oplus M_2$ becomes a left A-module when we define $a(m_1, m_2) = (am_1, am_2)$.

Suppose that M is a left A-module. A linear subspace N of M is a left A-submodule if $an \in N$ whenever $a \in A$ and $n \in N$. N is then a left A-module in a natural way.

If N_1 and N_2 are left A-submodules, then so is $N_1 + N_2$. The intersection of left A-submodules is again a left A-submodule. This means that if C is a non-empty subset of a left A-module M then there is a smallest left A-submodule containing C, which we denote by $[C]_A$. Then

$$[C]_A = \{a_1 c_1 + \cdots + a_k c_k : k \geq 0, a_i \in A, c_i \in C\}.$$

If $C = \{c_1, \ldots, c_n\}$, we write $[c_1, \ldots, c_n]_A$ for $[C]_A$: then

$$[c_1, \ldots, c_n]_A = \{a_1 c_1 + \cdots + a_n c_n : a_i \in A\}.$$

If M is a left A-module and $M = [c_1, \ldots, c_n]_A$, we say that M is a *finitely generated* left A-module. If $M = [c]_A = \{ac : a \in A\}$, then M is a *cyclic* left A-module, and c is called a *cyclic vector* for M.

Let us consider two examples. First, consider A as a left A-module. Then a subset J of A is a left A-submodule if and only if J is a left ideal in A. The left A-module A is cyclic, with cyclic vector 1. More generally, an element c of A is a cyclic vector if and only if c has a left inverse d in A (an element d such that $dc = 1$). For if $[c]_A = A$ then $1 \in [c]_A$, so that c must have a left inverse. Conversely, if d is a left inverse for c, then $a = adc \in [c]_A$, for any $a \in A$, and so c is a cyclic vector.

Secondly, suppose that M is a left A-module. Then M^k is also a left A-module. It can however also be considered as a left $\mathcal{M}_k(A)$-module: if $\alpha = (a_{ij}) \in \mathcal{M}_k(A)$ and $\mu = (m_j) \in M^k$, define $\alpha\mu$ by $(\alpha\mu)_i = \sum_{j=1}^{k} a_{ij} m_j$.

2.5 Module homomorphisms

Suppose that M_1 and M_2 are both left A-modules. Then a linear mapping T from M_1 to M_2 is called a *module homomorphism*, or *A-homomorphism*, if $T(am) = aT(m)$ for all $a \in A$ and all $m \in M_1$. A bijective A-homomorphism is called an *A-isomorphism*. The set of all A-homomorphisms from M_1 to M_2 is a linear subspace of $L(M_1, M_2)$, denoted by $\mathrm{Hom}_A(M_1, M_2)$. If M is a left A-module, we denote $\mathrm{Hom}_A(M, M)$ by $\mathrm{End}_A(M)$; it is a subalgebra of $L(M)$. Its elements are called *A-endomorphisms*.

For example, suppose that $N = N_1 \oplus \cdots \oplus N_n$ and $M = M_1 \oplus \cdots \oplus M_m$ are direct sums of left A-modules. Then a linear mapping $\theta : N \to M$

is an A-homomorphism if and only if there are A-homomorphisms θ_{ij} : $N_j \to M_i$ such that

$$\theta(x) = \left(\sum_{j=1}^{n} \theta_{ij} x_j \right)_{i=1}^{m}, \quad \text{for } x = (x_1, \ldots, x_n) \in N.$$

In particular, if $M_i = N_j = L$ then $\operatorname{Hom}_A(N, M)$ is isomorphic to the space $M_{mn}(\operatorname{End}_A(L))$ of $m \times n$ matrices with entries in $\operatorname{End}_A(L)$.

Proposition 2.5.1 *Suppose that A is a unital algebra. Then $\operatorname{End}_A(A)$ is naturally isomorphic as an algebra to A^{opp}.*

Proof Define $\phi(T) = T(1_A)$, for $T \in \operatorname{End}_A(A)$. If I is the identity in $\operatorname{End}_A(A)$, then $\phi(I) = 1_A = 1_{A^{opp}}$, and if $S, T \in \operatorname{End}_A(A)$ then

$$\phi(ST) = ST(1_A) = S(\phi(T)) = S(\phi(T)1_A) = \phi(T)\phi(S),$$

so that ϕ is an algebra homomorphism of $\operatorname{End}_A(A)$ into A^{opp}.

It remains to show that ϕ is bijective. Suppose that $\phi(T) = \phi(S)$. If $a \in A$ then

$$T(a) = T(a1_A) = aT(1_A) = aS(1_A) = S(a1_A) = S(a),$$

so that $T = S$. Thus ϕ is injective. Suppose that $c \in A^{opp}$. Let $\rho(c)(b) = bc$ (this is multiplication in A, not in A^{opp} !). Then $\rho(c)(ab) = abc = a(\rho(c)(b))$, so that $\rho(c) \in \operatorname{End}_A(A)$. Since $\phi(\rho(c)) = \rho(c)(1_A) = c$, ϕ is surjective. □

2.6 Simple modules

A left A-module M is said to be *simple* if it is not $\{0\}$ and has no left A-submodules other than $\{0\}$ and M itself. Thus A is simple if and only if the representation $\pi : A \to L(M)$ defined by $\pi(a)(x) = a.x$ is irreducible.

Proposition 2.6.1 *A left A-module M other than $\{0\}$ is simple if and only if every non-zero element of M is a cyclic vector.*

Proof If M is simple and $m \in M$ is non-zero then $[m]_A$ is a non-zero left A-submodule, and so must be M. Thus M is cyclic and m is a cyclic vector. Suppose conversely that every non-zero element of M is a cyclic vector. If n is a non-zero element of a left A-submodule N, then $M = [n]_A \subseteq N$, so that $N = M$; M is simple. □

Recall that if we consider an algebra A as a left A-module, then the left A-submodules of A are the left ideals of A. Thus a non-zero left ideal is a simple left A-module if and only if it is a minimal (non-zero) left ideal. A is a simple left A-module if and only if it has no left ideals other than $\{0\}$ and A itself. Note that this is stronger than being a simple algebra.

Proposition 2.6.2 *A unital algebra A is a simple left A-module if and only if A is a division algebra; that is, every element of A has a (two-sided) inverse.*

Proof If A is a division algebra, then every non-zero element of A has an inverse, and so is a cyclic vector. Thus A is simple, by Proposition 2.6.1. If conversely A is simple, and $a \neq 0$, then a is a cyclic vector, and so has a left inverse b. Then $b \neq 0$, and so b has a left inverse c. Then $c = c(ba) = (cb)a = a$, and so b is a right inverse of a, too. $\qquad\square$

Thus by Frobenius' theorem, $\mathrm{End}_A(M)$ is isomorphic either to $\mathbf{R}$, in which case the only A-endomorphisms of M are scalar multiples of the identity, or $\mathbf{C}$ or $\mathbf{H}$. All these possibilities can occur. Clearly $\mathrm{End}_{\mathbf{R}}(\mathbf{R}) \cong \mathbf{R}$. $\mathbf{C}$ can be considered as a simple real left $\mathbf{C}$-module and then $\mathrm{End}_{\mathbf{C}}(\mathbf{C}) \cong \mathbf{C}$. Similarly, $\mathbf{H}$ is a simple real left $\mathbf{H}$-module, and $\mathrm{End}_{\mathbf{H}}(\mathbf{H}) \cong \mathbf{H}^{opp} \cong \mathbf{H}$. But $\mathbf{C}$ and $\mathbf{H}$ are not simple $\mathbf{R}$-modules.

A-homomorphisms between simple left A-modules have an important property.

Theorem 2.6.1 (Schur's lemma) *Suppose that $T \in \mathrm{Hom}_A(M_1, M_2)$, where M_1 and M_2 are simple left A-modules. If $T \neq 0$ then T is an A-isomorphism of M_1 onto M_2, and T^{-1} ia also an A-isomorphism.*

Proof Suppose that $T \neq 0$. Then $N = \{x \in M_1 : T(x) = 0\}$ is a left A-submodule of M_1 which is not equal to M_1; thus $N = \{0\}$, and T is injective. Similarly, $T(M_1)$ is a non-zero left A-submodule of M_2, and so is equal to M_2. Thus T is an A-isomorphism.

Let T^{-1} be the inverse mapping. Then if $a \in A$ and $m \in M_2$,

$$T^{-1}(am) = T^{-1}(aTT^{-1}(m)) = T^{-1}(T(aT^{-1})(m)) = aT^{-1}(m),$$

so that T^{-1} is an A-isomorphism. $\qquad\square$

Exercise

1. If $z = x + iy \in \mathbf{C}$ and $h \in \mathbf{H}$, let $zh = xh + yih$. Show that this makes $\mathbf{H}$ a $\mathbf{C}$-module. Is it simple?

2.7 Semi-simple modules

A left A-module M is *semi-simple* if it can be written as a direct sum
$M = M_1 \oplus \cdots \oplus M_d$ of simple left A-modules.

Since each M_i is cyclic, a semi-simple left A-module is finitely generated.

Theorem 2.7.1 *Suppose that M is a finitely generated left A-module.
Then M is semi-simple if and only if M is the span of its simple left
A-submodules.*

Proof The condition is certainly necessary. Suppose that M is spanned
by its simple left A-submodules. If $M = [m_1, \ldots, m_k]_A$, then each m_i
is in the span of finitely many simple left A-submodules. Thus M is
spanned by finitely many simple left A-submodules. Let $\{M_1, \ldots, M_j\}$
be a minimal family of simple left A-submodules which spans M. Let
$L_i = [\cup\{M_k : k \neq i\}]_A$. Suppose that $M_i \cap L_i \neq \{0\}$. Then $M_i \cap L_i$ is
a left A-module contained in M_i; since M_i is simple, $M_i \cap L_i = M_i$, so
that $M_i \subseteq L_i$. Thus we can discard M_i, contradicting the minimality of
the family. Thus $M_i \cap L_i = \{0\}$ for each i, from which it follows that
$M = M_1 \oplus \cdots \oplus M_j$.   □

Proposition 2.7.1 *Suppose that $M = M_1 \oplus \cdots \oplus M_k$ is the direct sum
of simple left A-modules. If N is a non-zero simple submodule of M,
then N is isomorphic as a left A-module to M_i for some $1 \leq i \leq k$.*

Proof Let $j : N \to M$ be the inclusion mapping from N to M, and
let π_i denote the projection from M onto M_i for $1 \leq i \leq k$. Let x be a
non-zero element of N. There exists i such that $\pi_i(x) \neq 0$. Then $\pi_i \circ j$
is a non-zero module homomorphism of N into M_i. Since N and M_i
are simple, $\pi_i \circ j$ is a module isomorphism of N onto M_i, by Schur's
lemma.   □

In particular, if M is a finitely generated left **H**-module, there exist $m_1, \ldots, m_k$ in M such that every element m of M can be written uniquely as $m = h_1 m_1 + \cdots + h_k m_k$, where $h_1, \ldots, h_k$ are in **H**.
Then M has real dimension $4k$, so that k is uniquely determined; we set
$\dim_{\mathbf{H}} M = k$.

Theorem 2.7.2 *A finite-dimensional simple unital algebra A is semi-simple.*

Proof Suppose that J is a non-zero left ideal of minimal dimension.

Then J is a simple left A-module. Let

$$JA = \left\{ \sum_{i=1}^{k} j_i a_i : k \in \mathbf{N}, j_i \in J, a_i \in A \right\}.$$

Then JA is a non-zero ideal in A; since A is simple, $JA = A$. Thus there exist $j_1, \ldots, j_k$ in J and $a_1, \ldots, a_k$ in A such that $1_A = j_1 a_1 + \cdots + j_k a_k$. We may clearly suppose that each $j_i a_i \neq 0$, so that $J a_i$ is a non-zero left ideal. Now the mapping $j \to j a_i$ is a left A-module homomorphism of J onto $J a_i$. By Schur's lemma, it is an isomorphism, and so each $J a_i$ is also a simple left A-submodule. Since $J a_1, \ldots, J a_k$ span A, A is semi-simple. $\qquad\qquad\qquad\qquad\qquad\qquad\qquad\qquad\qquad\qquad\qquad\qquad$ $\square$

Corollary 2.7.1 (Wedderburn's theorem)   *If A is a finite-dimensional simple real unital algebra then there exists $k \in \mathbf{N}$ and a division algebra $\mathbf{D} = \mathbf{R}$, $\mathbf{C}$ or $\mathbf{H}$ such that $A \cong M_k(\mathbf{D})$. Thus $A \cong \mathrm{End}_{\mathbf{D}}(\mathbf{D}^k)$.*

Proof   We can write $A = J_1 \oplus \cdots \oplus J_k$ as a direct sum of simple left A-modules (minimal non-zero left ideals). We show that each J_i is module isomorphic to J_1. If not, rearranging the terms if necessary, we can write

$$A = (J_1 \oplus \cdots \oplus J_l) \oplus (J_{l+1} \oplus \cdots \oplus J_k) = B \oplus C,$$

where J_i is module isomorphic to J_1 if $1 \leq i \leq l$ and J_i is not module isomorphic to J_1 if $l + 1 \leq i \leq k$. Let π_B be the projection of A onto B and let π_C be the projection of A onto C. If $\theta \in \mathrm{Hom}_A(A)$, then $\pi_C(\theta(J_i)) = \pi_C(\theta(AJ_i)) = \pi_C(A\theta(J_i)) = A\pi_C(\theta(J_i))$, so that either $\pi_C(\theta(J_i)) = \{0\}$ or $\pi_C(\theta(J_i))$ is a submodule of C isomorphic to J_i. It therefore follows from Proposition 2.7.1 that if $1 \leq i \leq l$ then $\pi_C(\theta(J_i)) = \{0\}$, so that $\theta(J_i) \subseteq B$. Thus $\theta(B) \subseteq B$, and similarly $\theta(C) \subseteq C$. Since, for $a \in A$, the mapping $b \to ba : A \to A$ is a left A-module homomorphism, it follows that B and C are ideals in A, contradicting the simplicity of A.

By Proposition 2.6.2 there exists a division algebra $\mathbf{D}$ such that

$$\mathrm{Hom}_A(J_i, J_l) \cong \mathbf{D}, \quad \text{for } 1 \leq i, l \leq k.$$

Consequently

$$A^{opp} \cong \mathrm{End}_A(A) \cong \mathrm{End}_A(J_1 \oplus \cdots \oplus J_k) \cong M_k(\mathrm{Hom}_A(J_i, J_l)) \cong M_k(\mathbf{D}).$$

Since $M_k(\mathbf{D}) \cong (M_k(\mathbf{D}^{opp}))^{opp} \cong (M_k(\mathbf{D}))^{opp}$, the result follows. $\qquad$ $\square$

Corollary 2.7.2   *If a is an element of a finite-dimensional simple real* *unital algebra A, and if a has a left inverse or a right inverse, then a has a two-sided inverse.*

Proof   We can consider A as $\mathrm{End}_A(\mathbf{D}^k)$, and a as an element of the real vector space $L(\mathbf{D}^k)$ of real linear mappings from $\mathbf{D}^k$ to itself. Each of the conditions implies that a is a bijection, and so has an inverse a^{-1} in $L(\mathbf{D}^k)$. If $a^{-1}(x) = y$ and $d \in \mathbf{D}$ then $a(dy) = d(ay) = dx$, so that $a^{-1}(dx) = dy = d(a^{-1}(x)$; thus $a^{-1} \in \mathrm{End}_A(\mathbf{D}^k)$.   □

Are there other, essentially different, representations? The next theorem shows that the answer is 'no'.

Theorem 2.7.3   *Let* $\mathbf{D} = \mathbf{R}, \mathbf{C}$ *or* $\mathbf{H}$. *Suppose that W is a finite-dimensional real vector space, and that* $\pi : M_k(\mathbf{D}) \to L(W)$ *is a real unital representation. Then the mapping* $(\lambda, w) \to \pi(\lambda I)(w)$ *from* $\mathbf{D} \times W$ *to W makes W a left* $\mathbf{D}$-*module,* $\dim_{\mathbf{D}}(W) = rk$ *for some r, and there is a* π-*invariant decomposition*

$$W = W_1 \oplus \cdots \oplus W_r,$$

where $\dim_{\mathbf{D}}(W_s) = k$ *for each* $1 \leq s \leq r$. *For each* $1 \leq s \leq r$ *there is a basis* $(e_1, \ldots, e_k)$ *of* W_s *and an isomorphism* $\pi_s : A \to M_k(\mathbf{D})$ *such that*

$$\pi(A) \left(\sum_{j=1}^{k} x_j e_j \right) = \sum_{i=1}^{k} \left(\sum_{j=1}^{k} a_{ij} x_j \right) e_i, \quad \text{where } (a_{ij}) = \pi_s(A).$$

In other words, any irreducible unital representation of $M_k(\mathbf{D})$ is essentially the same as the natural representation of $M_k(\mathbf{D})$ as $\mathrm{Hom}_{\mathbf{D}}(\mathbf{D}^k)$, and any finite-dimensional representation is a direct sum of irreducible representations.

Proof   It is immediate that the mapping $(\lambda, w) \to \pi(\lambda I)(w)$ from $\mathbf{D} \times W$ to W makes W a left $\mathbf{D}$-module.

Let the matrix $E(i, j)$ be defined by setting

$$E(i,j)_{i,j} = 1 \text{ and } E(i,j)_{l,m} = 0 \text{ otherwise, for } 1 \leq i, j, l, m \leq k,$$

and let $T_{ij} = \pi(E(i,j))$. Since $E(j,j)$ is a projection, so is T_{jj}. Further,

$$T_{ii}T_{jj} = 0 \text{ for } i \neq j, \text{ and } I = T_{11} + \cdots + T_{kk}.$$

Let $U_j = T_{jj}(W)$. Then $W = U_1 \oplus \cdots \oplus U_k$. Since $T_{ii}T_{ij}T_{jj} = T_{ij}$, T_{ij} is an isomorphism of U_j onto U_i, with inverse T_{ji}. Let $(f_1, \ldots, f_r)$ be a basis for U_1. We use this to express W as the direct sum $W_1 \oplus \cdots \oplus W_r$ of r π-invariant subspaces, each of dimension k. Suppose that $1 \leq s \leq r$.

Let $e_j^{(s)} = T_{j1}(f_s)$, for $1 \leq j \leq k$. Then $e_j^{(s)} \in U_j$, and $(e_1^{(s)}, \ldots, e_k^{(s)})$ is a basis for $W_s = \mathrm{span}(e_1^{(s)}, \ldots, e_k^{(s)})$.

Since $T_{ij}(e_j^{(s)}) = e_i^{(s)}$ and $T_{ij}(e_k^{(s)}) = 0$ for $k \neq j$, W_s is π-invariant, and

$$\pi(A) \left(\sum_{j=1}^{k} x_j e_j^{(s)} \right) = \sum_{i=1}^{k} \left(\sum_{j=1}^{k} a_{ij} x_j \right) e_i^s \quad \text{for some } (a_{ij}).$$

The mapping $\pi_s : M_k(\mathbf{D}) \to M_k(\mathbf{D})$ defined by $\pi_s(A) = (a_{ij})$ is then an isomorphism of A onto $M_k(\mathbf{D})$, by Schur's lemma.

Finally, $W = W_1 \oplus \cdots \oplus W_r$. $\qquad\square$

3

Multilinear algebra

To each finite-dimensional vector space E equipped with a non-singular quadratic form, there is associated a universal Clifford algebra A. If $\dim E = d$ then $\dim A = 2^d$; each time we increase the dimension of the vector space by 1, the dimension of the algebra doubles. This suggests strongly and correctly that the algebra should be constructed as a tensor product. Many mathematicians feel uncomfortable with tensor product spaces, since they are constructed as a quotient of an infinite-dimensional space by an infinite-dimensional subspace. Here we avoid this, by making systematic use of the duality theory of finite-dimensional vector spaces. In the same way, we use duality to construct the exterior algebra of a finite-dimensional vector space (which is a particular example of a Clifford algebra) and to construct its symmetric algebra.

In this chapter, K will denote either the field $\mathbf{R}$ of real numbers or the field $\mathbf{C}$ of complex numbers.

3.1 Multilinear mappings

Suppose that $E_1, \ldots, E_k$ and F are vector spaces over K. A mapping $T : E_1 \times \cdots \times E_k \to F$ is *multilinear*, or *k-linear*, if it is linear in each variable:

$$T(x_1, \ldots, x_{j-1}, \alpha x_j + \beta y_j, x_{j+1}, \ldots, x_k) =$$
$$\alpha T(x_1, \ldots, x_{j-1}, x_j, x_{j+1}, \ldots, x_k) + \beta T(x_1, \ldots, x_{j-1}, y_j, x_{j+1}, \ldots, x_k),$$

for $\alpha, \beta \in K$, $x_j, y_j \in E_j$ and $1 \le j \le k$. Under pointwise addition, the k-linear mappings from $E_1 \times \cdots \times E_k$ into F form a vector space, which is denoted by $M(E_1, \ldots, E_k; F)$. When $E_1 = \cdots = E_k = E$, we write $M^k(E, F)$ for $M(E_1, \ldots, E_k; F)$. We write $M(E_1, \ldots, E_k)$ for

$M(E_1, \ldots, E_k; K)$, and $M^k(E)$ for $M^k(E, K)$. Elements of $M(E_1, \ldots, E_k)$ are called *multilinear forms*, or *k-linear forms*.

A 2-linear mapping is called a *bilinear* mapping. The vector space of bilinear mappings from $E_1 \times E_2$ into F is denoted by $B(E_1, E_2; F)$. We write $B(E_1, E_2)$ for $B(E_1, E_2; K)$ and $B(E)$ for $B(E, E)$. Elements of $B(E_1, E_2)$ are called *bilinear forms*.

Proposition 3.1.1 *If $b \in B(E_1, E_2; F)$, $e_1 \in E_1$ and $e_2 \in E_2$, let $l_b(e_1) : E_2 \to F$ be defined by $l_b(e_1)(e_2) = b(e_1, e_2)$. Then $l_b(e_1) \in L(E_2, F)$, and the mapping $l_b : E_1 \to L(E_2, F)$ is linear. The mapping $l : b \to l_b$ is a linear isomorphism of $B((E_1, E_2; F)$ onto $L(E_1, L(E_2, F))$. Similarly, the mapping r defined by $r_b(e_2)(e_1) = b(e_1, e_2)$ is a linear isomorphism of $B((E_1, E_2; F)$ onto $L(E_2, L(E_1, F))$.*

Proof Straightforward verification. □

Corollary 3.1.1 $\dim B(E_1, E_2; F) = \dim E_1 . \dim E_2 . \dim F$.

Similarly, $M(E_1, \ldots, E_k; F)$ is isomorphic to $M(E_1, \ldots, E_{k-1}; L(E_k, F))$, and an inductive argument shows that

$$\dim M(E_1, \ldots, E_k; F) = \left(\prod_{j=1}^{k} \dim E_j \right) . \dim F.$$

Corollary 3.1.2 $B(E_1, E_2) \cong L(E_1, E_2') \cong L(E_2, E_1')$, *and*

$$\dim B(E_1, E_2) = \dim E_1 . \dim E_2.$$

Proposition 3.1.2 *If $B \in B(E_1, E_2)$ then $\mathrm{rank}(l_b) = \mathrm{rank}(r_b)$.*

Proof If $x_2 \in E_2$, then $x_2 \in (l_b(E_1))^\perp$ if and only if $b(x_1, x_2) = 0$ for all $x_1 \in E_1$, and this happens if and only if $r_b(x_2) = 0$. Thus $(l_b(E_1))^\perp = N(r_b)$, and so

$$\mathrm{rank}(l_b) = \dim E_2 - \dim(l_b(E_1))^\perp$$
$$= \dim E_2 - \dim N(r_b) = \mathrm{rank}(r_b),$$

by the rank-nullity formula. □

If $b \in B(E_1, E_2)$, we define the *rank* of b to be the rank of the element l_b of $L(E_1, E_2')$. b is *non-singular* if $\mathrm{rank}(l_b) = \dim E_1 = \dim E_2$. Thus b is non-singular if and only if l_b and r_b are invertible.

Exercises

1. Prove Proposition 3.1.1.
2. Suppose that $(e_1, \ldots, e_d)$ is a basis for E_1, with dual basis $(\phi_1, \ldots, \phi_d)$, and that $(f_1, \ldots, f_c)$ is a basis for E_2, with dual basis $(\psi_1, \ldots, \psi_c)$. Suppose that b is a bilinear form on $E_1 \times E_2$, and that $b_{ij} = b(e_i, f_j)$ for $1 \leq i \leq d$, $1 \leq j \leq c$. Explain how the matrix (b_{ij}) is related to the matrices which represent l_b and r_b.

3.2 Tensor products

We now use duality to define the tensor product of vector spaces.

Suppose that $(x_1, \ldots, x_k) \in E_1 \times \cdots \times E_k$. The *evaluation mapping* $m \to m(x_1, \ldots, x_k)$ from $M(E_1, \ldots, E_k)$ into K is a linear functional on $M(E_1, \ldots, E_k)$. We denote it by $x_1 \otimes \cdots \otimes x_k$, and call it an *elementary tensor*. We denote the linear span of all such elementary tensors by $E_1 \otimes \cdots \otimes E_k$. $E_1 \otimes \cdots \otimes E_k$ is the *tensor product* of $(E_1, \ldots, E_k)$. In fact, the elementary tensors span the dual $M'(E_1, \ldots, E_k)$ of $M(E_1, \ldots, E_k)$.

Proposition 3.2.1 $E_1 \otimes \cdots \otimes E_k = M'(E_1, \ldots, E_k)$.

Proof If $m \in (E_1 \otimes \cdots \otimes E_k)^\perp$, then $m(x_1, \ldots, x_k) = 0$ for all $(x_1, \ldots, x_k) \in E_1 \times \cdots \times E_k$, and so $m = 0$. Thus

$$E_1 \otimes \cdots \otimes E_k = (E_1 \otimes \cdots \otimes E_k)^{\perp\perp} = M'(E_1, \ldots, E_k).$$

$\square$

Corollary 3.2.1 $\dim(E_1 \otimes \cdots \otimes E_k) = \prod_{j=1}^{k}(\dim E_j)$.

We can obtain a basis for $E_1 \otimes \cdots \otimes E_k$ by taking a basis for each of the spaces $E_1, \ldots, E_k$, and taking all tensor products of the form $e_1 \otimes \cdots \otimes e_k$, where each e_j is a member of the appropriate basis.

The tensor product has the following universal mapping property, which enables us to replace a multilinear mapping by a linear one.

Proposition 3.2.2   *The mapping* $\otimes^k : E_1 \times \cdots \times E_k \to E_1 \otimes \cdots \otimes E_k$ *defined by* $\otimes^k(x_1, \ldots, x_k) = x_1 \otimes \cdots \otimes x_k$ *is k-linear.*

If $m \in M(E_1, \ldots E_k; F)$ *there exists a unique linear mapping* $L(m) \in L(E_1 \otimes \cdots \otimes E_k) \to F$ *such that* $m = L(m) \circ \otimes^k$; *that is,*

$$m(x_1, \ldots x_k) = L(m)(x_1 \otimes \cdots \otimes x_k) \quad for \ (x_1, \ldots x_k) \in E_1 \times \cdots \times E_k.$$

The mapping $L : m \to L(m)$ is an isomorphism of $M(E_1, \ldots, E_k; F)$ onto $L(E_1 \otimes \cdots \otimes E_k, F)$.

Thus we have the following diagram.

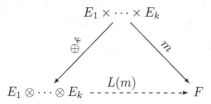

Proof Since

$$(x_1 \otimes \cdots \otimes (\alpha x_j + \beta y_j) \otimes \cdots \otimes x_k)(m)$$
$$= m(x_1, \ldots, (\alpha x_j + \beta y_j), \ldots, x_k)$$
$$= \alpha m(x_1, \ldots, x_j, \ldots, x_k) + \beta m(x_1, \ldots, y_j, \ldots, x_k)$$
$$= \alpha(x_1 \otimes \cdots \otimes x_j \otimes \cdots \otimes x_k)(m)$$
$$+ \beta(x_1 \otimes \cdots \otimes y_j \otimes \cdots \otimes x_k)(m),$$

$\otimes^k$ is k-linear.

The mapping $T \to T \circ \otimes^k$ is a linear mapping from $L(E_1 \otimes \cdots \otimes E_k, F)$ into $M(E_1, \ldots, E_k; F)$. If $T \circ \otimes^k = 0$, then $T(x_1 \otimes \cdots \otimes x_k) = 0$ for all elementary tensors $x_1 \otimes \cdots \otimes x_k$. Since these span $E_1 \otimes \cdots \otimes E_k$, $T = 0$. Thus the mapping $T \to T \circ \otimes^k$ is injective. Since $L(E_1 \otimes \cdots E_k, F)$ and $M(E_1, \ldots, E_k; F)$ have the same dimension, it is also surjective; this gives the result. $\qquad\square$

Corollary 3.2.2 *(i) There is a unique isomorphism $\kappa : E \otimes F \to L(F', E)$ such that*

$$\kappa(x \otimes y)(\phi) = \phi(y)x \quad \text{for } x \in E, y \in F, \phi \in F'.$$

(ii) There is a unique isomorphism $\lambda : E \otimes F' \to L(F, E)$ such that

$$\lambda(x \otimes \phi)(y) = \phi(y)x \quad \text{for } x \in E, y \in F, \phi \in F'.$$

Proof (i) If $(x, y) \in E \times F$, let $m(x, y)(\phi) = \phi(y)x$, for $\phi \in F'$. Then $m(x, y) \in L(F', E)$, and m is a bilinear mapping from $E \times F$ into $L(F', E)$. Let $\kappa = L(m)$ be the corresponding linear mapping from $E \otimes F$ into $L(F', E)$. We must show that κ is an isomorphism. Let $(e_1, \ldots, e_m)$ be a basis for E, let $(f_1, \ldots, f_n)$ be a basis for F, and let $(\phi_1, \ldots, \phi_n)$ be the basis for F' dual to $(f_1, \ldots, f_n)$. If $t \in E \otimes F$, we can write t as $\sum_{i=1}^{m} \sum_{j=1}^{n} t_{ij} e_i \otimes f_j$. Then $\kappa(t)(\phi_j) = \sum_{i=1}^{m} t_{ij} e_i$, so that if $\kappa(t) = 0$

then $t_{ij} = 0$ for all i and j, and $t = 0$. Thus κ is injective. Since $E \otimes F$ and $L(F', E)$ have the same dimension, it follows that κ is an isomorphism.
(ii) This follows, since $F'' \cong F$. ☐

We define the *rank* of an element t of $E \otimes F$ to be the rank of $\kappa(t)$. Properties of the rank are obtained in Exercise 1.

Corollary 3.2.3 $(E_1 \otimes \cdots \otimes E_k) \otimes (E_{k+1} \otimes \cdots \otimes E_l)$ *is naturally isomorphic to* $E_1 \otimes \cdots \otimes E_l$.

Proof For

$$(E_1 \otimes \cdots \otimes E_k) \otimes (E_{k+1} \otimes \cdots \otimes E_l)$$
$$= B'(E_1 \otimes \cdots \otimes E_k, E_{k+1} \otimes \cdots \otimes E_l)$$
$$\cong B'(M'(E_1, \ldots, E_k), M'(E_{k+1}, \ldots, E_l))$$
$$\cong M'(E_1, \ldots, E_l) = E_1 \otimes \cdots \otimes E_l.$$

☐

We denote the tensor product of k copies of a vector space E by $\otimes^k E$. Then it follows from the preceding corollary that the infinite direct sum

$$\otimes^* E = K \oplus E \oplus (E \otimes E) \oplus \cdots \oplus (\otimes^k E) \oplus \cdots$$

is an infinite-dimensional associative algebra, the *tensor algebra* of E. In particular, $(\otimes^k E) \otimes (\otimes^l E) \subseteq \otimes^{k+l} E$. The tensor algebra has the following universal property.

Theorem 3.2.1 *Suppose that* $T : E \to A$ *is a linear mapping of* E *into a unital algebra* A. *Then* T *extends uniquely to a unital algebra homomorphism* $\tilde{T} : \otimes^* E \to A$.

Thus we have the following diagram.

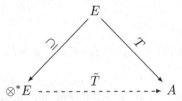

Proof Let $(e_1, \ldots, e_d)$ be a basis for E. If $e_{i_1} \otimes \cdots \otimes e_{i_k}$ is a corresponding basic vector in $\otimes^k(E)$, let

$$\tilde{T}(e_{i_1} \otimes \cdots \otimes e_{i_k}) = T(e_{i_1}) \otimes \cdots \otimes T(e_{i_k})$$

and extend by linearity to $\otimes^k E$, and then to $\otimes^* E$. $\tilde{T}$ is then a unital algebra homomorphism. Since $e_1, \ldots, e_d$ generate $\otimes^* E$, the extension is unique. $\qquad\square$

We can also consider the tensor product of linear operators. Suppose that $T_j \in L(E_j, F_j)$ for $1 \leq j \leq k$. If $m \in M(F_1, \ldots, F_k)$, let

$$T(m)(x_1, \ldots, x_k) = m(T_1(x_1), \ldots, T_k(x_k))$$

for $(x_1, \ldots, x_k) \in E_1 \otimes \cdots \otimes E_k$. Then $T(m) \in M(E_1, \ldots, E_k)$, and $T \in L(M(F_1, \ldots, F_k), M(E_1, \ldots, E_k))$. Let T' be its transpose, so that $T' \in L(E_1 \otimes \cdots \otimes E_k, F_1 \otimes \cdots \otimes F_k)$. The mapping $(T_1, \ldots, T_k) \to T'$ from $L(E_1, F_1) \times \cdots \times L(E_k, F_k)$ to $L(E_1 \otimes \cdots \otimes E_k, F_1 \otimes \cdots \otimes F_k)$ is k-linear, and so there exists a linear mapping

$$j : L(E_1, F_1) \otimes \cdots \otimes L(E_k, F_k) \to L(E_1 \otimes \cdots \otimes E_k, F_1 \otimes \cdots \otimes F_k)$$

such that $j(T_1 \otimes \cdots \otimes T_k) = T'$. By slight abuse of notation, we denote $j(T_1 \otimes \cdots \otimes T_k)$ by $T_1 \otimes \cdots \otimes T_k$. Thus

$$(T_1 \otimes \cdots \otimes T_k)(x_1 \otimes \cdots \otimes x_k) = T_1(x_1) \otimes \cdots \otimes T_k(x_k).$$

The representation of tensor products of linear mappings by matrices can become complicated. If $I_1, \ldots, I_k$ are the index sets for bases of $E_1, \ldots, E_k$, and $J_1, \ldots, J_k$ are the index sets for bases of $F_1, \ldots, F_k$, then $T_1 \otimes \cdots \otimes T_k$ is represented by a $I_1 \times \cdots \times I_k$ by $J_1 \times \cdots \times J_k$ matrix. We shall use the following convention, in the case where $k = 2$. If $S \otimes T$ maps $E_1 \otimes E_2$ into $F_1 \otimes F_2$ and if S,T also denote matrices representing the mappings, then we represent $S \otimes T$ by the matrix

$$\begin{bmatrix} St_{11} & \cdots & St_{1n} \\ \vdots & \ddots & \vdots \\ St_{m1} & \cdots & St_{mn} \end{bmatrix},$$

where in each place St_{ij} is a multiple of the matrix S.

For example, if $E_1 = E_2 = F_1 = F_2 = \mathbf{R}^2$ and U and Q are given by the matrices

$$U = \begin{bmatrix} 1 & 0 \\ 0 & -1 \end{bmatrix} \text{ and } Q = \begin{bmatrix} 0 & 1 \\ 1 & 0 \end{bmatrix},$$

then $U \otimes Q$ is represented by

$$\begin{bmatrix} 0 & U \\ U & 0 \end{bmatrix} = \begin{bmatrix} 0 & 0 & 1 & 0 \\ 0 & 0 & 0 & -1 \\ 1 & 0 & 0 & 0 \\ 0 & -1 & 0 & 0 \end{bmatrix}$$

and $Q \otimes U$ is represented by

$$\begin{bmatrix} Q & 0 \\ 0 & -Q \end{bmatrix} = \begin{bmatrix} 0 & 1 & 0 & 0 \\ 1 & 0 & 0 & 0 \\ 0 & 0 & 0 & -1 \\ 0 & 0 & -1 & 0 \end{bmatrix}.$$

There are several other ways of defining the tensor product of vector spaces.

We can consider the infinite-dimensional vector space of all K-valued functions on $E_1 \times \cdots \times E_k$ which take the value 0 on all but finitely many points, and define the tensor product as a quotient of this by an appropriate infinite-dimensional subspace. (This is the only practicable method when considering tensor products of more general objects, such as modules.) With this definition, it is necessary to establish existence and uniqueness properties carefully.

At the other extreme, we can consider a basis B_j for each space E_j, and define the tensor product as the space of all K-valued functions on $B_1 \times \cdots \times B_k$. We set $e_{i_1}^{(1)} \otimes \cdots \otimes e_{i_k}^{(k)}$ at the indicator function of $\{(e_{i_1}^{(1)}, \ldots, e_{i_k}^{(k)})\}$, define $\otimes^k (e_{i_1}^{(1)}, \ldots, e_{i_k}^{(k)}) = e_{i_1}^{(1)} \otimes \cdots \otimes e_{i_k}^{(k)}$, and extend by multilinearity. This has the disadvantage that it is co-ordinate based, and it is necessary to determine the effect of changes of bases.

Exercises

1. Suppose that $t \in E \otimes F$ has rank r. Show that t can be written as $t = x_1 \otimes y_1 + \cdots + x_r \otimes y_r$.

   Show that in such a representation each of the sequences $(x_1, \ldots, x_r)$ and $(y_1, \ldots, y_r)$ is linearly independent.

   Conversely show that if $t = x_1 \otimes y_1 + \cdots + x_r \otimes y_r$, where each of the sequences $(x_1, \ldots, x_r)$ and $(y_1, \ldots, y_r)$ is linearly independent, then t has rank r.

2. In the example above, show that $(U \otimes Q)(Q \otimes U) = (Q \otimes U)(U \otimes Q)$, and show that $((U \otimes Q)(Q \otimes U))^2 = -I$.

3.3 The trace

Suppose that E is a d-dimensional vector space over K, with basis $(e_1, \ldots, e_d)$ and dual basis $(\phi_1, \ldots, \phi_d)$. Let us consider the tensor product $E \otimes E'$, which has as basis $\{e_i \otimes \phi_j : 1 \le i, j \le d\}$. The mapping

$(x, \psi) \to \psi(x) : E \times E' \to K$ is bilinear, and so there exists $t \in (E \otimes E')'$ such that $t(x \otimes \psi) = \psi(x)$ for all $x \in E, \psi \in E'$. $t(x \otimes \psi)$ is the *contraction* of the elementary tensor $x \otimes \psi$.

Recall (Corollary 3.2.2 (ii)) that there is a natural isomorphism λ of $E \otimes E'$ onto $L(E)$. If $T \in L(E)$, we set $\tau(T) = t(\lambda^{-1}(T))$. τ is the *trace* of T. If T is represented by the matrix (t_{ij}), then

$$\tau(T) = t\left(\sum_{i,j} t_{ij} e_i \otimes \phi_j\right) = \sum_{i,j} t_{ij}\phi_j(e_i) = \sum_{i=1}^{d} t_{ii}.$$

Note though that $\tau(I) = d$; it is therefore often more convenient to work with the *normalized trace* $\tau_n = \tau/\dim E$.

We can use the trace, or the normalized trace, to define a duality between spaces of operators. Suppose that F is another vector space, with basis $(f_1, \ldots, f_c)$. If $R \in L(E, F)$ is represented by the matrix (r_{ij}) and $S \in L(F, E)$ is represented by the matrix (s_{ji}), then $SR \in L(E)$ is represented by the matrix $(\sum_{j=1}^{c} r_{ij}s_{jk})$ and $RS \in L(F)$ is represented by the matrix $(\sum_{i=1}^{d} s_{ji}r_{ik})$, so that $\tau(RS) = \tau(SR) = \sum_{i,j} r_{ij}s_{ji}$. Thus if we set $b(R, S) = \tau(RS) = \tau(SR)$ then b is a bilinear form on $L(E, F) \times L(F, E)$. In particular, the trace defines a bilinear form, the *trace form*, b, on $L(E)$.

Suppose now that A is a finite-dimensional unital algebra. The left regular representation l is an algebra isomorphism of A into $L(A)$. We define $\tau(a) = \tau(l_a)$ and $\tau_n(a) = \tau_n(l_a)$; τ is the *trace* and τ_n the *normalized trace* on A. Then $\tau(ab) = \tau(ba)$ and $\tau_n(ab) = \tau_n(ba)$, for $a, b \in A$. Since l_1 is the identity mapping on A, $\tau_n(1) = 1$. Elements of the null space of τ_n are called *pure* elements of A, and elements of span(1) are called *scalars*. If $a \in A$, let $Pu(x) = x - \tau_n(x)1$. Then Pu is a projection of A onto the subspace $Pu(A)$ of pure elements of A, and $A = \text{span}(1) \oplus Pu(A)$. This definition is consistent with the definition of 'pure quaternion' defined in Section 2.3, since if $x = a1 + bi + cj + dk \in \mathbf{H}$ then, with respect to the basis $(1, i, j, k)$, l_x is represented by the matrix

$$\begin{bmatrix} a & -b & -c & -d \\ b & a & -d & c \\ c & d & a & -b \\ d & -c & b & a \end{bmatrix},$$

so that $\tau(x) = 4a$ and $\tau_n(x) = a$.

Exercises

1. Show that the the mapping $(S,T) \to \tau(ST)$ is a non-singular bilinear form on $L(E,F) \times L(F,E)$.

2. Suppose that $T_1 \in L(E_1)$ and $T_2 \in L(E_2)$. Show that

$$\tau(T_1 \otimes T_2) = \tau(T_1)\tau(T_2).$$

3. Suppose that ϕ is a linear functional on $L(E)$ which satisfies

$$\phi(I) = 1 \quad \text{and} \quad \phi(ST) = \phi(TS) \text{ for all } S,T \in L(E).$$

Show that $\phi = \tau_n$.

4. Give an example of a finite-dimensional unital algebra A for which there is more than one linear functional ϕ on A for which

$$\phi(I) = 1 \quad \text{and} \quad \phi(ab) = \phi(ba) \text{ for all } a,b \in A.$$

5. Suppose that $T \in L(E)$. Let $f(t) = \det(I + tT)$. Show that

$$\tau(T) = \frac{df}{dt}(0).$$

6. Suppose that A is a two-dimensional real unital algebra. Show that A is a super-algebra, with direct sum decomposition $\mathbf{R}.1 \oplus Pu(A)$. Suppose that x is a non-zero pure element of A. Show that if $x^2 > 0$ then $A \cong \mathbf{R}^2$ and if $x^2 < 0$ then $A \cong \mathbf{C}$. Describe A when $x^2 = 0$.

3.4 Alternating mappings and the exterior algebra

Throughout this section, we suppose that E is a d-dimensional vector space over K, with basis $(e_1, \ldots, e_d)$ and dual basis $(\phi_1, \ldots, \phi_d)$ and that F is a vector space over K.

A k-linear mapping $a : E^k \to F$ is *alternating* if $a(x_1, \ldots, x_k) = 0$ whenever there exist distinct i,j for which $x_i = x_j$.

Proposition 3.4.1 *Suppose that $m \in M^k(E,F)$. The following are equivalent.*

(i) m is alternating.

(ii) $m(x_1, \ldots, x_k) = \epsilon(\sigma)m(x_{\sigma(1)}, \ldots, x_{\sigma(k)})$ for $\sigma \in \Sigma_k$ (where $\epsilon(\sigma)$ is the signature of the permutation σ).

(iii) If i,j are distinct elements of $\{1, \ldots, k\}$, and if $x_i' = x_j$, $x_j' = x_i$ and $x_l' = x_l$ for all other indices l, then $m(x_1, \ldots, x_k) = -m(x_1', \ldots, x_k')$.

Proof (ii) implies (iii), since (iii) is just (ii) applied to a transposition, and (iii) implies (ii), since any permutation can be written as a product of transpositions. (iii) clearly implies (i). Let us show that (i) implies (iii). To keep the terminology simple, let us suppose that $i = 1$ and $j = 2$; the argument clearly applies to the other cases:

$$
\begin{aligned}
0 &= m(x_1 + x_2, x_1 + x_2, x_3, \dots, x_k) \\
&= m(x_1, x_1, x_3, \dots, x_k) + m(x_1, x_2, x_3, \dots, x_k) \\
&\quad + m(x_2, x_1, x_3, \dots, x_k) + m(x_2, x_2, x_3, \dots, x_k) \\
&= m(x_1, x_2, x_3, \dots, x_k) + m(x_2, x_1, x_3, \dots, x_k).
\end{aligned}
$$

$\square$

We denote the set of alternating k-linear mappings of E^k into F by $A^k(E, F)$; it is clearly a linear subspace of $M^k(E; F)$. We write $A^k(E)$ for $A^k(E, K)$, and denote the dual of $A^k(E)$ by $\bigwedge^k(E)$. $\bigwedge^k(E)$ is the *kth exterior product* of E.

Suppose that $(x_1, \dots, x_k) \in E^k$. We denote the evaluation mapping $a \to a(x_1, \dots, x_k)$ from $A^k(E)$ to K by $x_1 \wedge \dots \wedge x_k$; $x_1 \wedge \dots \wedge x_k$ is the *alternating product*, or *wedge product*, of $x_1, \dots, x_d$. It is a linear functional on $A^k(E)$, and so is an element of $\bigwedge^k(E)$. We denote the mapping $(x_1, \dots, x_k) \to x_1 \wedge \dots \wedge x_k : E^k \to \bigwedge^k(E)$ by $\wedge^k$; $\wedge^k$ is an alternating k-linear mapping.

Proposition 3.4.2

$$
\begin{aligned}
\bigwedge{}^k(E) &= \mathrm{span}\{x_1 \wedge \dots \wedge x_k : (x_1, \dots, x_k) \in E^k\} \\
&= \mathrm{span}\{e_{j_1} \wedge \dots \wedge e_{j_k} : 1 \le j_1 < \dots < j_k \le d\}.
\end{aligned}
$$

Proof Suppose that $a \in \mathrm{span}\{x_1 \wedge \dots \wedge x_k : (x_1, \dots, x_k) \in E^k\}^\perp$. Then $a(x_1, \dots x_k) = (x_1 \wedge \dots \wedge x_k)(a) = 0$ for all $(x_1, \dots, x_k) \in E^k$, so that $a = 0$. Thus

$$
\bigwedge{}^k(E) = \mathrm{span}\{x_1 \wedge \dots \wedge x_k : (x_1, \dots, x_k) \in E^k\}.
$$

Expanding $x_1, \dots, x_k$ in terms of the basis, and using the fact that $\wedge^k$ is an alternating k-linear mapping, it follows that

$$
(x_1, \dots, x_k) \in \mathrm{span}\{e_{j_1} \wedge \dots \wedge e_{j_k} : 1 \le j_1 < \dots < j_k \le d\}.
$$

$\square$

$A^k(E)$ is a linear subspace of $M^k(E)$. On the other hand, the mapping

$P_k : M^k(E) \to A^k(E)$ defined by

$$P_k(m)(x_1, \ldots, x_k) = \frac{1}{k!} \sum_{\sigma \in \Sigma_k} \epsilon(\sigma) m(x_{\sigma(1)}, \ldots, x_{\sigma(k)})$$

is a natural projection of $M^k(E)$ onto $A^k(E)$. We denote the transposed inclusion mapping from $\bigwedge^k(E)$ into $\otimes^k(E)$ by J_k:

$$J_k(x_1 \wedge \ldots \wedge x_k) = \frac{1}{k!} \sum_{\sigma \in \Sigma_k} \epsilon(\sigma) x_{\sigma(1)} \otimes \cdots \otimes x_{\sigma(k)}.$$

Proposition 3.4.3 $\{e_{j_1} \wedge \ldots \wedge e_{j_k} : 1 \le j_1 < \cdots < j_k \le d\}$ *is a basis for* $\bigwedge^k(E)$.

Proof We need to show that the set

$$\{e_{j_1} \wedge \ldots \wedge e_{j_k} : 1 \le j_1 < \cdots < j_k \le d\}$$

is linearly independent.

Suppose that $\boldsymbol{j} = (j_1, \ldots, j_k)$, with $1 \le j_1 < \cdots < j_k \le d$. Let

$$m_{\boldsymbol{j}}(x_1, \ldots, x_k) = \phi_{j_1}(x_1) \ldots \phi_{j_k}(x_k).$$

Then $m_{\boldsymbol{j}} \in M^k(E)$; let $a_{\boldsymbol{j}} = P_k(m_{\boldsymbol{j}})$. Then $a_{\boldsymbol{j}} \in A^k(E)$, and $a_{\boldsymbol{j}}(e_{j_1}, \ldots, e_{j_k}) = 1/k!$, while

$$a_{\boldsymbol{j}}(e_{l_1}, \ldots, e_{l_k}) = 0 \text{ if } l_1 < \cdots < l_k \text{ and } \{l_1, \ldots, l_k\} \neq \{j_1, \ldots, j_k\}.$$

Suppose now that

$$u = \sum \{u_{j_1, \ldots, j_k}(e_{j_1} \wedge \ldots \wedge e_{j_k}) : 1 \le j_1 < \cdots < j_k \le d\} = 0.$$

Then $u_{j_1, \ldots, j_k} = k! a_{\boldsymbol{j}}(u) = 0$, for each $\boldsymbol{j} = (j_1, \ldots, j_k)$, and so the set $\{e_{j_1} \wedge \ldots \wedge e_{j_k} : 1 \le j_1 < \cdots < j_k \le d\}$ is linearly independent. $\square$

Corollary 3.4.1

$$\dim \left(\bigwedge {}^k(E) \right) = \binom{d}{k} = \frac{d!}{k!(d-k)!}.$$

In particular, $\bigwedge^d(E)$ is one-dimensional, and $\bigwedge^k(E) = \{0\}$ for $k > d$. If $x_1 \wedge \ldots \wedge x_k \in \bigwedge^k(E)$ and $x_{k+1} \wedge \ldots \wedge x_l \in \bigwedge^{l-k}(E)$ then, as for tensor products, $x_1 \wedge \ldots \wedge x_l \in \bigwedge^l(E)$. Extending by bilinearity, we see that if we set

$$\bigwedge {}^*(E) = K \oplus E \oplus \cdots \oplus \bigwedge {}^k(E) \oplus \cdots \oplus \bigwedge {}^d(E),$$

then $\bigwedge^*(E)$ is a unital associative algebra of dimension 2^d; we denote multiplication by $\wedge$. This is the *exterior algebra* $\bigwedge^*(E)$ of E. It becomes a super-algebra, when we set

$$\bigwedge{}^+(E) = \mathrm{span}\{\bigwedge{}^k(E) : k \text{ even}\},$$

$$\bigwedge{}^-(E) = \mathrm{span}\{\bigwedge{}^k(E) : k \text{ odd}\}.$$

If $T \in L(E, F)$ and $1 \leq k \leq d$ then T defines an element $T^{(k)}$: $A^k(F) \to A^k(E)$, defined by

$$T^{(k)}(a)(x_1, \ldots, x_k) = a(T(x_1), \ldots, T(x_k)).$$

We denote the transpose mapping in $L(\bigwedge^k(E), \bigwedge^k(F))$ by $\wedge^k(T)$. Then $\wedge^k(T)(x_1 \wedge \ldots \wedge x_k) = T(x_1) \wedge \ldots \wedge T(x_k)$. In particular, if $E = F$ and $k = d$ then $\wedge^d(T)$ is a linear mapping from the one-dimensional space $\bigwedge^d(E)$ into itself, and so there exists an element $\det T$ of K such that $\wedge^d(T)(x_1 \wedge \ldots \wedge x_d) = (\det T)(x_1 \wedge \ldots \wedge x_d)$. $\det T$ is the *determinant* of T.

If $a, b \in \bigwedge^*(E)$, let $l_a(b) = a \wedge b$. Then the mapping $l : a \to l_a$ is the *left regular representation* of the algebra $\bigwedge^*(E)$ in $L(\bigwedge^*(E))$; it is a unital isomorphism of $\bigwedge^*(E)$ onto a subalgebra of $L(\bigwedge^*(E))$. In particular, if $x \in E$, then we denote the operator l_x by m_x: m_x is called a *creation operator*. m_x maps $\bigwedge^k(E)$ into $\bigwedge^{k+1}(E)$.

We can also define *annihilation operators*. Before we do so, let us introduce some terminology. If $(x_1, \ldots, x_{k+1})$ is a sequence with $k + 1$ terms, let $(x_1, \ldots, \check{x}_j, \ldots, x_{k+1})$ denote the sequence with k terms obtained by deleting the term x_j. Similarly $(x_1, \ldots, \check{x}_i, \ldots, \check{x}_j, \ldots, x_{k+1})$ denotes the sequence with $k - 1$ terms obtained by deleting the terms x_i and x_j. We use a similar notation for wedge products.

Suppose that $\phi \in E'$ and $a \in A^k(E)$. Define $l_\phi(a) \in M^{k+1}(E)$ by setting $l_\phi(a)(x_1, \ldots, x_{k+1}) = \phi(x_1)a(x_2, \ldots, x_{k+1})$. $l_\phi(a)$ is not alternating; we set $P_\phi(a) = P_{k+1}l_\phi(a)$, so that

$$P_\phi(a)(x_1, \ldots, x_{k+1}) = \frac{1}{(k+1)!} \sum_{\sigma \in \Sigma_{k+1}} \epsilon(\sigma)\phi(x_{\sigma(1)})a(x_{\sigma(2)}, \ldots, x_{\sigma(k+1)}).$$

An alternative, and useful, expression is given by

$$P_\phi(a)(x_1, \ldots, x_{k+1}) = \frac{1}{k+1} \sum_{j=1}^{k+1} (-1)^{j-1}\phi(x_j)a(x_1, \ldots, \check{x}_j, \ldots, x_{k+1}).$$

Then $P_\phi(a) \in A^{k+1}(E)$, and $P_\phi \in L(A^k(E), A^{k+1}(E))$.

Let $\delta_\phi : \bigwedge^{k+1}(E) \to \bigwedge^k(E)$ be the transpose mapping. We set

$\delta_\phi(\lambda) = 0$, for $\lambda \in K$. Letting k vary, we consider δ_ϕ as an element of $L(\bigwedge^*(E))$. δ_ϕ is the *annihilation operator* defined by ϕ. Easy calculations then show that

$$\delta_\phi(x_1 \wedge \cdots \wedge x_k)$$

$$= \frac{1}{k} \sum_{j=1}^{k} (-1)^{j-1} \phi(x_j)(x_1 \wedge \cdots \wedge \breve{x}_j \wedge \cdots \wedge x_k)$$

$$= \frac{1}{k!} \sum_{\sigma \in \Sigma_k} \epsilon(\sigma) \phi(x_{\sigma(1)})(x_{\sigma(2)} \wedge \cdots \wedge x_{\sigma(k)}).$$

Theorem 3.4.1 $m_x^2 = \delta_\phi^2 = 0$, and $m_x \delta_\phi + \delta_\phi m_x = \phi(x)I$.

Proof Since $x \wedge x = 0$, $m_x^2 = 0$. If $a \in K$ then $\delta_\phi^2(a) = 0$. If $a \in A^k(E)$ then

$$(k+1)(k+2)P_\phi^2(a)(x_1, \ldots x_{k+2})$$

$$= (k+1) \sum_{j=1}^{k+2} (-1)^{j-1} \phi(x_j) P_\phi(a)(x_1, \ldots, \breve{x}_j, \ldots, x_{k+2})$$

$$= \sum_{1 \le i < j \le k+2} (-1)^{j-1}(-1)^{i-1} \phi(x_j)\phi(x_i) a(x_1, \ldots, \breve{x}_i, \ldots, \breve{x}_j, \ldots, x_{k+2})$$

$$\quad + \sum_{1 \le j < i \le k+2} (-1)^{j-1}(-1)^{i-2} \phi(x_j)\phi(x_i) a(x_1, \ldots, \breve{x}_j, \ldots, \breve{x}_i, \ldots, x_{k+2})$$

$$= 0,$$

so that $\delta_\phi^2 = (P_\phi^2)' = 0$. Next,

$$m_x \delta_\phi(x_1 \wedge \cdots \wedge x_k)$$

$$= m_x \left(\frac{1}{k} \sum_{j=1}^{k} (-1)^{j-1} \phi(x_j)(x_1 \wedge \cdots \wedge \breve{x}_j \wedge \cdots \wedge x_k) \right)$$

$$= \frac{1}{k} \sum_{j=1}^{k} (-1)^{j-1} \phi(x_j)(x \wedge x_1 \wedge \cdots \wedge \breve{x}_j \wedge \cdots \wedge x_k),$$

while

$$\delta_\phi m_x(x_1 \wedge \cdots \wedge x_k) = \delta_\phi(x \wedge x_1 \wedge \cdots \wedge x_k)$$

$$= \phi(x)(x_1 \wedge \cdots \wedge x_k)$$

$$\quad + \frac{1}{k} \sum_{j=1}^{k} (-1)^j \phi(x_j)(x \wedge x_1 \wedge \cdots \wedge \breve{x}_j \wedge \cdots \wedge x_k)$$

so that $m_x \delta_\phi + \delta_\phi m_x = \phi(x)I$. $\qquad\qquad\qquad\qquad\qquad\qquad$ □

Exercises

1. Let $j_k : A^k(E) \to M^k(E)$ be the inclusion mapping, and let $q_k : \otimes^k(E) \to \bigwedge^k(E)$ be the transpose mapping. Show that

$$
\begin{aligned}
x_1 \wedge \ldots \wedge x_k &= q_k(x_1 \otimes \cdots \otimes x_k) \\
&= \epsilon(\sigma) q_k(x_{\sigma(1)} \otimes \cdots \otimes x_{\sigma(k)}), \text{ for } \sigma \in \Sigma_k, \\
&= \frac{1}{k!} \sum_{\sigma \in \Sigma_k} \epsilon(\sigma) q_k(x_{\sigma(1)} \otimes \cdots \otimes x_{\sigma(k)}).
\end{aligned}
$$

2. Combine the q_k, by varying k, to obtain a surjective linear mapping $q : \otimes^*(E) \to \bigwedge^*(E)$. Show that this is a unital algebra homomorphism. Show that the kernel of this homomorphism is the ideal J generated by the set $\{x \otimes x : x \in E\}$. (This provides another construction of the exterior algebra.)

3. Suppose that $x \in E$ and that $x \neq 0$. Show that m_x has rank 2^{d-1}, and that the null-space of m_x is the same as the image of m_x. [Hint: Consider a basis $(e_1, \ldots, e_d)$ for E with $x = e_1$.]

4. Suppose that $\phi \in E'$ and that $\phi \neq 0$. Show that δ_ϕ has rank 2^{d-1}, and that the null-space of δ_ϕ is the same as the image of δ_ϕ. [Hint: Consider a basis $(e_1, \ldots, e_d)$ for E with $\phi(e_1) = 1$ and $\phi(e_j) = 0$ otherwise.]

5. Suppose that $(e_1, \ldots, e_d)$ is a basis for E. Let $e_\Omega = e_1 \wedge \ldots \wedge e_d$. Show that $\mathbf{R}e_\Omega$ is a one-dimensional ideal in $\bigwedge^*(E)$.

3.5 The symmetric tensor algebra

We can also create symmetric tensor products. We shall not need them later, and so we describe the construction and state the basic results, leaving the details to the reader. Suppose again that E is a d-dimensional vector space over K, with basis $(e_1, \ldots, e_d)$.

A k-linear mapping $s : E^k \to F$ is *symmetric* if

$$
s(x_1, \ldots, x_k) = s(x_{\sigma(1)}, \ldots, x_{\sigma(k)}) \text{ for each } \sigma \in \Sigma_k.
$$

The set $S^k(E, F)$ of symmetric k-linear mapping $s : E^k \to F$ is a linear subspace of $M^k(E; F)$. We denote $S^k(E, K)$ by $S^k(E)$. We denote the evaluation mapping $s \to s(x_1, \ldots, x_k)$ from $S^k(E)$ to K by

$x_1 \otimes_s \ldots \otimes_s x_k$. Then $x_1 \otimes_s \ldots \otimes_s x_k \in (S^k(E))'$. We denote the linear span of the *symmetric tensor products* $x_1 \otimes_s \ldots \otimes_s x_k$ by $\otimes_s^k(E)$. We denote the mapping $(x_1, \ldots, x_k) \to x_1 \otimes_s \ldots \otimes_s x_k$ by $\otimes_s^k$; $\otimes_s^k$ is a symmetric k-linear mapping.

Proposition 3.5.1

$$(S^k(E))' = \otimes_s^k(E) = \operatorname{span}\{e_{j_1} \otimes_s \ldots \otimes_s e_{j_k} : 1 \leq j_1 \leq \cdots \leq j_k \leq d\}.$$

$S^k(E)$ is a linear subspace of $M^k(E)$. On the other hand, the mapping $P_k^{(s)} : M^k(E) \to S^k(E)$ defined by

$$P_k^{(s)}(m)(x_1, \ldots, x_k) = \frac{1}{k!} \sum_{\sigma \in \Sigma_k} m(x_{\sigma(1)}, \ldots, x_{\sigma(k)})$$

is a natural projection of $M^k(E)$ onto $S^k(E)$. We denote the transposed inclusion mapping from $\otimes_s^k(E)$ into $\otimes^k(E)$ by $J_k^{(s)}$:

$$J_k^{(s)}(x_1 \otimes_s \ldots \otimes_s x_k) = \frac{1}{k!} \sum_{\sigma \in \Sigma_k} x_{\sigma(1)} \otimes \cdots \otimes x_{\sigma(k)}.$$

Proposition 3.5.2 $\{e_{j_1} \otimes_s \ldots \otimes_s e_{j_k} : 1 \leq j_1 \leq \cdots \leq j_k \leq d\}$ *is a basis for* $\otimes_s^k(E)$.

Corollary 3.5.1

$$\dim(\otimes_s(E)) = \binom{n+k-1}{k-1} = \frac{(n+k-1)!}{n!(k-1)!}.$$

If $x_1 \otimes_s \ldots \otimes_s x_k \in \bigotimes_s^k(E)$ and $x_{k+1} \otimes_s \ldots \otimes_s x_l \in \bigotimes_s^{l-k}(E)$ then, as for tensor products, $x_1 \otimes_s \ldots \otimes_s x_l \in \bigotimes_s^l(E)$. Extending by bilinearity, we see that if we set

$$\bigotimes_s^*(E) = K \oplus E \oplus \cdots \oplus \bigotimes_s^k(E) \oplus \cdots,$$

then $\bigotimes_s^*(E)$ is a commutative infinite-dimensional unital associative algebra. This is the *symmetric tensor algebra* of E.

If $T \in L(E, F)$ and $1 \leq k \leq d$ then T defines a linear mapping $T^{(k)} : S^k(F) \to S^k(E)$, defined by

$$T^{(k)}(s)(x_1, \ldots x_k) = s(T(x_1), \ldots, T(x_k)).$$

We denote the transpose mapping in $L(\bigotimes_s^k E, \bigotimes_s^k F)$ by $\otimes_s^k(T)$. Then $\otimes_s^k(T)(x_1 \otimes_s \ldots \otimes_s x_k) = T(x_1) \otimes_s \ldots \otimes_s T(x_k)$.

If $a, b \in \bigotimes_s^*(E)$, let $l_a(b) = a \otimes_s b$. Then the mapping $l : a \to l_a$ is the *left regular representation* of $\bigotimes_s^*(E)$ in $L(\bigotimes_s^*(E))$; it is a unital isomorphism of $\bigotimes_s^*(E)$ onto a subalgebra of $L(\bigotimes_s^*(E))$. In particular, if $x \in E$, then the operator l_x is called a *creation operator*, and is denoted by m_x. m_x is an injective linear mapping of $\bigotimes_s^k(E)$ into $\bigotimes_s^{k+1}(E)$.

We can also define *annihilation operators*. Suppose that $\phi \in E'$. If $s \in S^k(E)$, let $\lambda_\phi(s)(x_1, \dots, x_{k+1}) = \phi(x_1)s(x_2, \dots, x_{k+1})$. Then $\lambda_\phi(a) \in M^{k+1}(E)$; let $P_\phi^{(s)}(a) = P_{k+1}^{(s)}\lambda_\phi(a)$. Then $P_\phi^{(s)}(s)(a) \in S^{k+1}(E)$, and $P_\phi^{(s)} \in L(S^k(E), S^{k+1}(E))$. Let $\delta_\phi : \bigotimes_s^{k+1} E \to \bigotimes_s^k E$ be the transpose mapping. Letting k vary, we consider δ_ϕ as an element of $L(\bigotimes_s^* E)$. δ_ϕ is the *annihilation operator* defined by ϕ. Easy calculations then show that

$$\delta_\phi(x_1 \otimes_s \cdots \otimes_s x_{k+1})$$

$$= \sum_{j=1}^k \phi(x_j)(x_2, \otimes_s \cdots \otimes_s x_{j-1} \otimes_s x_1 \otimes_s x_{j+1} \otimes_s \cdots \otimes_s x_k)$$

$$= \frac{1}{k!} \sum_{\sigma \in \Sigma_{k+1}} \phi(x_{\sigma(1)})(x_{\sigma(2)} \otimes_s \cdots \otimes_s x_{\sigma(k+1)}).$$

The pattern of this section is intended to show the close parallel between exterior algebras and symmetric tensor algebras. But the differences are very important! In quantum physics, an exterior algebra is called a *fermionic Fock space*, and a symmetric tensor algebra is called a *bosonic Fock space*; the differences between them reflect the fundamental differences between fermions and bosons. The notions of creation operators and annihilation operators also come from quantum physics.

Exercises

1. Prove Corollary 3.5.1.

2. Suppose that b is a bilinear form on $E \times E$. Show that b is symmetric if and only if the linear mappings $l_b : E \to E'$ and $r_b : E \to E'$ are equal.

3. Let $j_k : S^k(E) \to M_k(E)$ be the inclusion mapping, and let $q_k : \otimes^k(E) \to \otimes_s^k(E)$ be the transpose mapping. Extend this, by varying k, to a surjective linear mapping $q : \otimes^*(E) \to \otimes_s^*(E)$. Show that this is a unital algebra homomorphism. Show that the kernel is the ideal generated by the set $\{x \otimes y - y \otimes x : x, y \in E\}$. (This provides another construction of the symmetric tensor algebra.)

4. Let A be a unital algebra, and let J be the ideal generated by all

elements of the form $ab - ba$. Show that the quotient algebra A/J is commutative. Show that the construction of the previous exercise is an example of this.

5. State and prove a universal mapping theorem for commutative algebras and symmetric tensor algebras.

3.6 Tensor products of algebras

Suppose that A and B are unital algebras. We can form the vector space tensor product $A \otimes B$, and turn it into an associative algebra by setting

$$(a_1 \otimes b_1)(a_2 \otimes b_2) = a_1 a_2 \otimes b_1 b_2.$$

This is a unital associative algebra (the identity is $1_A \otimes 1_B$). The elements $\{a \otimes 1_B : a \in A\}$ form a subalgebra isomorphic to A, and similarly the elements $\{1_A \otimes b : b \in B\}$ form a subalgebra isomorphic to B, and so we consider A and B as subalgebras of $A \otimes B$. Further,

$$ab = (a \otimes 1_B)(1_A \otimes b) = a \otimes b = (1_A \otimes b)(a \otimes 1_B) = ba,$$

so that elements taken from A and B commute. Let us establish some basic properties of the tensor product of two algebras.

Theorem 3.6.1 *Suppose that A and B are finite-dimensional unital algebras. Then $Z(A \otimes B) = Z(A) \otimes Z(B)$.*

Proof It follows from the definition of multiplication that $Z(A) \otimes Z(B) \subseteq Z(A \otimes B)$. Suppose that $t \in Z(A \otimes B)$ has rank r. Then $t = a_1 \otimes b_1 + \cdots + a_r \otimes b_r$, where each of the sequences $(a_1, \ldots, a_r)$ and $(b_1, \ldots, b_r)$ is linearly independent (Section 3.2, Exercise 1). If $a \in A$ then

$$\sum_{j=1}^{r} a a_j \otimes b_j = (a \otimes 1_B)t = t(a \otimes 1_B) = \sum_{j=1}^{r} a_j a \otimes b_j,$$

so that each a_j is in $Z(A)$. Similarly, each b_j is in $Z(B)$, so that $t \in Z(A) \otimes Z(B)$. Thus $Z(A \otimes B) \subseteq Z(A) \otimes Z(B)$. $\qquad \square$

Theorem 3.6.2 *Suppose that A and B are finite-dimensional simple unital algebras, and that A is central. Then $A \otimes B$ is simple.*

Proof Suppose that I is a non-zero ideal in $A \otimes B$. We must show that $I = A \otimes B$, or that $1_A \otimes 1_B \in I$. Let $t = \sum_{j=1}^{r} a_j \otimes b_j$ be a non-zero element of I of minimal rank r. Then each of the sequences $(a_1, \ldots, a_r)$

and $(b_1, \ldots, b_r)$ is linearly independent. Since A is simple, there exist elements $c_1, \ldots, c_k$ and $d_1, \ldots, d_k$ of A such that $1_A = \sum_{i=1}^{k} c_i a_1 d_i$. Let $a'_j = \sum_{i=1}^{k} c_i a_j d_i$, for $2 \leq j \leq k$. Then

$$s = \sum_{i=1}^{k} (c_1 \otimes 1_B) t(d_i \otimes 1_B) = 1 \otimes b_1 + \sum_{j=2}^{r} a'_j \otimes b_j$$

is in I. Since the sequence $(b_1, \ldots, b_r)$ is linearly independent, $s \neq 0$, and therefore s has rank r, and the sequence $(1, a'_2, \ldots, a'_r)$ is linearly independent. If $a \in A$, then

$$u = (a \otimes 1_B)s - s(a \otimes 1_B) = \sum_{j=2}^{r} (a a'_j - a'_j a) \otimes b_j \in I.$$

It follows from the minimality of r that $u = 0$, so that $a'_j \in Z(A)$, for $2 \leq j \leq r$. But $\dim Z(A) = 1$, and the sequence $(1, a'_2, \ldots, a'_r)$ is linearly independent. Consequently, $r = 1$ and $s = 1 \otimes b_1$. But there exist elements $e_1, \ldots, e_l$ and $f_1, \ldots, f_k$ of B such that $1_B = \sum_{i=1}^{l} e_i b_1 f_i$, and so

$$1_A \otimes 1_B = \sum_{i=1}^{l} (1_A \otimes e_i)(1 \otimes b_1)(1_A \otimes f_i) \in I.$$

$\square$

How do we recognize that an algebra is isomorphic to the tensor product of two algebras? We say that two subalgebras F and G of an algebra C *commute* if $fg = gf$ for $f \in F$, $g \in G$.

Proposition 3.6.1 *Suppose that F and G are subalgebras of a finite-dimensional unital algebra C, which commute, and which generate C. Then there is a unique algebra homomorphism ϕ of $F \otimes G$ onto C which satisfies $\phi(f \otimes g) = fg$ for $(f, g) \in F \times G$.*

Proof The conditions are necessary, by the remarks above. Suppose that they are satisfied. Define a mapping $\theta : F \times G \to C$ by setting $\theta(f, g) = fg$. Then θ is a bilinear mapping, and so there exists a unique linear mapping $\phi : F \otimes G \to C$ such that $\phi(f \otimes g) = fg$. Since

$$\phi(f_1 \otimes g_1)\phi(f_2 \otimes g_2) = (f_1 g_1)(f_2 g_2) = (f_1 f_2)(g_1 g_2)$$
$$= \phi(f_1 f_2 \otimes g_1 g_2),$$

ϕ is an algebra homomorphism. Since $F \cup G$ generates C, ϕ is surjective.

$\square$

When is this homomorphism an isomorphism?

Corollary 3.6.1   *If F is a central simple algebra and G is a simple algebra then ϕ is an isomorphism of $F \otimes G$ onto C.*

Proof   Since $F \otimes G$ is simple, by Theorem 3.6.2, either ϕ is injective, when ϕ is an isomorphism of $F \otimes G$ onto C, or $\phi(F \otimes G) = \text{span}(1_C)$. In the latter case, C is one dimensional, and the result is trivially true. $\square$

Corollary 3.6.2   *The mapping ϕ is an isomorphism if and only if* $\dim C = (\dim F)(\dim G)$.

Proof   Since $\dim F \otimes G = (\dim F)(\dim G)$, ϕ is an isomorphism if and only if $\dim C = (\dim F)(\dim G)$. $\square$

We use this last corollary to prove the following.

Proposition 3.6.2   *Suppose that A is a real finite-dimensional unital algebra with identity I, and consider $\mathbf{C}$ and $\mathbf{H}$ as real algebras. Consider $\mathbf{R}^d$ as a real algebra, with co-ordinatewise multiplication:*

$$(x_1, \dots, x_d)(y_1, \dots, y_d) = (x_1 y_1, \dots, x_d y_d).$$

(i) $\mathbf{R}^d \otimes A \cong A^d$.
(ii) $M_d(\mathbf{R}) \otimes A \cong M_d(A)$.
(iii) $\mathbf{C} \otimes \mathbf{C} \cong \mathbf{C} \oplus \mathbf{C} \cong \mathbf{R}^2 \otimes \mathbf{C}$.
(iv) $\mathbf{C} \otimes \mathbf{H} \cong M_2(\mathbf{C})$.
(v) $\mathbf{H} \otimes \mathbf{H} \cong M_2(\mathbf{R}) \otimes M_2(\mathbf{R}) \cong M_4(\mathbf{R})$.

Proof   (i) Let I be the identity in A. Take

$$F = \{(x_1 I, \dots, x_d I) : x_j \in \mathbf{R}, \text{ for } 1 \le j \le d\}$$
$$\text{and } G = (a, \dots a) : a \in A.$$

Then F and G are subalgebras of A^d which commute and generate A^d, $F \cong \mathbf{R}^d$, $G \cong A$ and $\dim A^d = \dim F. \dim G$.
  (ii) Take

$$F = \{(x_{ij} I) : x_{ij} \in \mathbf{R}, \text{ for } 1 \le i, j \le d\}$$
$$\text{and } G = \{\text{diag}(a, \dots, a) : a \in A\}.$$

Then F and G are subalgebras of $M_d(A)$ which commute and generate $M_d(A)$, $F \cong M(\mathbf{R}^d)$, $G \cong A$ and $\dim M_d(A) = \dim F. \dim G$.
  (iii) Take $F = \{(z, z) : z \in \mathbf{C}\}$ and $G = \{(w, \bar{w}) : w \in \mathbf{C}\}$. Then F and

G are subalgebras of $\mathbf{C} \oplus \mathbf{C}$ which commute and are isomorphic to $\mathbf{C}$, and $\dim \mathbf{C} \oplus \mathbf{C} = 4 = (\dim F)(\dim G)$. Since

$$(1,0) = \tfrac{1}{2}((i,i)(-i,i) + (1,1)),$$
$$(i,0) = \tfrac{1}{2}((i,i) + (i,-i)),$$
$$(0,1) = \tfrac{1}{2}((i,i)(i,-i) + (1,1)),$$
$$(0,i) = \tfrac{1}{2}((i,i) + (-i,i)),$$

$\mathbf{C} \oplus \mathbf{C}$ is generated by $F \cup G$. Thus $\mathbf{C} \otimes \mathbf{C} \cong \mathbf{C} \oplus \mathbf{C}$. By (i), $\mathbf{R}^2 \otimes \mathbf{C} \cong \mathbf{C} \oplus \mathbf{C}$.

(iv) Take $F = \mathrm{diag}\{(z,z) : z \in \mathbf{C}\}$, and let $G = H$, the subalgebra of $M_2(\mathbf{C})$ defined in Section 2.3: G is spanned by the associate Pauli matrices τ_0, τ_1, τ_2 and τ_3. Then $F \cong \mathbf{C}$, $G \cong \mathbf{H}$, F and G commute, and $\dim M_2(\mathbf{C}) = 8 = (\dim F)(\dim G)$. The subalgebra of $M_2(\mathbf{C})$ generated by $F \cup G$ contains the matrices $\pm I$, $\pm J$, $\pm U$, $\pm Q$, and so is equal to $M_2(\mathbf{C})$.

(v) Define linear mappings $\theta_F : \mathbf{H} \to M_2(\mathbf{R}) \otimes M_2(\mathbf{R})$ and $\theta_G : \mathbf{H} \to M_2(\mathbf{R}) \otimes M_2(\mathbf{R})$ by setting

$$\theta_F(1) = I \otimes I, \quad \theta_F(\boldsymbol{i}) = Q \otimes J, \quad \theta_F(\boldsymbol{j}) = -J \otimes I, \quad \theta_F(\boldsymbol{k}) = U \otimes J,$$

$$\theta_G(1) = I \otimes I, \quad \theta_G(\boldsymbol{i}) = J \otimes Q, \quad \theta_G(\boldsymbol{j}) = -I \otimes J, \quad \theta_F(\boldsymbol{k}) = J \otimes U,$$

and extending by linearity. Then θ_F and θ_G are isomorphisms of $\mathbf{H}$ onto subalgebras F and G of $M_2(\mathbf{R}) \otimes M_2(\mathbf{R})$, which commute, and satisfy $\dim(M_2(\mathbf{R}) \otimes M_2(\mathbf{R})) = (\dim F)(\dim G)$. It is straightforward to verify that

$$\{\theta_F(u)\theta_G(v) : u, v \in \{1, \mathbf{i}, \mathbf{j}, \mathbf{k}\}\}$$

is a basis for $M_2(\mathbf{R}) \otimes M_2(\mathbf{R})$, so that $M_2(\mathbf{R}) \otimes M_2(\mathbf{R})$ is generated by $F \cup G$. $\qquad\square$

Exercise

1. Let $C = \mathbf{R}^3$, with multiplication defined by

$$(x_1, x_2, x_3)(y_1, y_2, y_3) = (x_1 y_1, x_1 y_2 + x_2 y_1, x_1 y_3 + x_3 y_1).$$

Show that C is a unital algebra, with identity $(1,0,0)$. Let $F = \{(x_1, x_2, 0) : x_1, x_2 \in \mathbf{R}\}$ and let $G = \{(x_1, 0, x_3) : x_1, x_2 \in \mathbf{R}\}$. Show that F and G are subalgebras of C which commute, that C is generated by $F \cup G$ and that $F \cap G = \mathrm{span}(I)$. Show that C is not isomorphic to $F \otimes G$.

3.7 Tensor products of super-algebras

When A and B are super-algebras there is another multiplication, the *graded tensor multiplication*, which takes the grading into account. If $A = A^+ \oplus A^-$ and $B = B^+ \oplus B^-$ are super-algebras, we define graded tensor multiplication by setting

$$(a_1 \otimes b_1)_g(a_2 \otimes b_2) = -a_1 a_2 \otimes b_1 b_2 \quad \text{if } b_1 \in B^- \text{ and } a_2 \in A^-,$$
$$= a_1 a_2 \otimes b_1 b_2 \quad \text{otherwise,}$$

for a_1, a_2 homogeneous in A and b_1, b_2 homogeneous in B, and then extending by linearity. We must check that this defines an associative multiplication. It is straightforward but tedious to check that if a_1, a_2, a_3 are homogeneous in A and b_1, b_2, b_3 are homogeneous in B then

$$[(a_1 \otimes b_1)_g(a_2 \otimes b_2)]_g(a_3 \otimes b_3) = (a_1 \otimes b_1)_g[(a_2 \otimes b_2)_g(a_3 \otimes b_3)]$$
$$= -a_1 a_2 a_3 \otimes b_1 b_2 b_3$$

in the six following cases

b_1	$-$	$-$	$-$	$-$	$+$	$+$
a_2	$-$	$-$	$-$	$+$	$+$	$-$
b_2	$-$	$-$	$+$	$+$	$-$	$-$
a_3	$-$	$+$	$+$	$-$	$-$	$-$

while

$$[(a_1 \otimes b_1)_g(a_2 \otimes b_2)]_g(a_3 \otimes b_3) = (a_1 \otimes b_1)_g[(a_2 \otimes b_2)_g(a_3 \otimes b_3)]$$
$$= a_1 a_2 a_3 \otimes b_1 b_2 b_3$$

in the remaining ten cases. We write $A \otimes_g B$ for $A \otimes B$ with this law of multiplication, and write $a \otimes_g b$ for the elementary tensors in it. Then $1_A \otimes_g 1_B$ is the identity for $A \otimes_g B$, and the mappings $a \to a \otimes_g 1_B$ and $b \to 1_A \otimes_g b$ are algebra monomorphisms of A and B into $A \otimes_g B$. We identify A and B with their images in $A \otimes_g B$.

Note that

$$(a^+ + a^-)_g(b^+ + b^-) = a^+ \otimes_g b^+ + a^+ \otimes_g b^- + a^- \otimes_g b^+ + a^- \otimes_g b^-$$
$$(b^+ + b^-)_g(a^+ + a^-) = a^+ \otimes_g b^+ + a^+ \otimes_g b^- + a^- \otimes_g b^+ - a^- \otimes_g b^-.$$

$A \otimes_g B$ is then a super-algebra, with

$$(A \otimes_g B)^+ = (A^+ \otimes_g B^+) \oplus (A^- \otimes_g B^-)$$
$$(A \otimes_g B)^- = (A^+ \otimes_g B^-) \oplus (A^- \otimes_g B^+).$$

Note that $A \otimes_g B \cong B \otimes_g A$. It is also straightforward but tedious to check associativity: $(A \otimes_g B) \otimes_g C = A \otimes_g (B \otimes_g C)$.

Let us give an example. As we have seen, the real algebra $\mathbf{C}$ is a super-algebra. Let

$$A = \mathbf{C} = \mathbf{R} \oplus i\mathbf{R} = A^+ \oplus A^-, \quad B = \mathbf{C} = \mathbf{R} \oplus j\mathbf{R} = B^+ \oplus B^-,$$

where to avoid confusion, we replace i by j in the definition of B. Let $\mathbf{H} = A \otimes_g B$, and write $\boldsymbol{i} = i \otimes_g 1_B$, $\boldsymbol{j} = 1_A \otimes_g j$ and $\boldsymbol{k} = \boldsymbol{i}_g \boldsymbol{j} = (i \otimes_g I_B)_g (I_A \otimes_g j)$. Then

$$\boldsymbol{i}_g^2 = i^2 \otimes_g I_B = -1,$$
$$\boldsymbol{j}_g^2 = I_A \otimes_g j^2 = -1,$$
$$\boldsymbol{i}_g \boldsymbol{j} = (i \otimes_g 1_B)_g (1_A \otimes_g j) = i \otimes_g j = \boldsymbol{k},$$
$$\boldsymbol{j}_g \boldsymbol{i} = (1_A \otimes_g j)_g (i \otimes_g 1_B) = -i \otimes_g j = -\boldsymbol{k},$$

and

$$\boldsymbol{k}_g^2 = (i \otimes_g j)_g (i \otimes_g j) = -(i^2 \otimes_g j^2) = -1$$
$$\boldsymbol{j}_g \boldsymbol{k} = (1 \otimes_g j)_g (i \otimes_g j) = -(i \otimes_g j^2) = \boldsymbol{i} = -\boldsymbol{j}_g \boldsymbol{k}$$
$$\boldsymbol{k}_g \boldsymbol{i} = (i \otimes_g j)_g (i \otimes_g 1) = -(i^2 \otimes_g j) = \boldsymbol{j} = -\boldsymbol{i}_g \boldsymbol{k}.$$

Thus $\mathbf{C} \otimes_g \mathbf{C}$ is isomorphic to the algebra $\mathbf{H}$ of quaternions.

Exercises

1. Show that if E and F are vector spaces then

$$\left(\bigwedge{}^{*}(E) \right) \otimes_g \left(\bigwedge{}^{*}(F) \right) \cong \bigwedge{}^{*}(E \oplus F).$$

2. Suppose that $A = A^+ \oplus A^-$ is a super-algebra. Show that $\tau_n(a^-) = 0$ for all $a^- \in A^-$, and that if τ_n^+ is the normalized trace on A^+ then $\tau_n^+(a^+) = \tau_n(a^+)$ for all $a^+ \in A^+$.

PART TWO

QUADRATIC FORMS AND CLIFFORD ALGEBRAS

4

Quadratic forms

Readers are probably familiar with Euclidean spaces. These are finite-dimensional real vector spaces, equipped with a positive definite quadratic form, which defines an inner product, a norm and the Euclidean metric.

Here we consider more general quadratic forms, such as arise in the theory of relativity. The theory of such forms is considerably more complicated than the theory of positive definite forms.

4.1 Real quadratic forms

Suppose that E is a real vector space. A real-valued function q on E is called a *quadratic form* on E if there exists a symmetric bilinear form b on E such that $q(x) = b(x, x)$, for all $x \in E$, and a vector space E equipped with a quadratic form q is called a *quadratic space* (E, q). Thus each symmetric bilinear form on E defines a quadratic form on E. The set $Q(E)$ of quadratic forms on E is a linear subspace of the vector space of all real-valued functions on E.

Proposition 4.1.1 *Distinct symmetric bilinear forms define distinct quadratic forms.*

Proof Suppose that $q(x) = b(x, x)$. Then

$$q(x + y) = b(x + y, x + y) = b(x, x) + b(x, y) + b(y, x) + b(y, y)$$
$$= q(x) + q(y) + 2b(x, y),$$

so that $b(x, y) = \frac{1}{2}(q(x + y) - q(x) - q(y))$. A similar calculation shows that $b(x, y) = \frac{1}{4}(q(x + y) - q(x - y))$. (These equations are called the *polarization formulae*.) Thus q determines b uniquely. $\square$

We call b the bilinear form *associated* with q. Since b is symmetric, the left mapping $l_b : E \to E'$ and the right mapping $r_b : E \to E'$ are the same. We define the *rank* of q to be the rank of b, and say that q is *non-singular* if the associated bilinear form is non-singular. A quadratic space is said to be *regular* if the quadratic form is non-singular. A linear subspace of a quadratic space is a quadratic space in a natural way.

We say that a quadratic form q is *positive definite* if $q(v) > 0$ for all non-zero v, and *positive semi-definite* if $q(v) \geq 0$ for all v. *Negative definite* and *negative semi-definite* quadratic forms are defined similarly. When q is positive definite, the associated bilinear form is called an *inner product* on E, and E is called an *inner-product space*, or a *Euclidean* space. The inner product is frequently written as $\langle x, y \rangle$, rather than $b(x, y)$. It is clear that an inner-product space is non-singular.

If F is a linear subspace of a quadratic space, then we denote the restriction of q to F by q. Then (F, q) ia a quadratic space.

Proposition 4.1.2 *Suppose that (E, q) is a regular quadratic space. Then every linear subspace of E is regular if and only if q is positive definite or negative definite.*

Proof It is clear that if q is positive definite, then its restriction to a linear subspace is positive definite, and is therefore non-singular. A similar argument applies to negative definite quadratic forms. Suppose that q is neither positive definite nor negative definite. Then there exist $c, d \in E$ with $q(c) > 0$ and $q(d) < 0$. Then c and d are linearly independent, and the quadratic polynomial

$$Q(\lambda) = q(\lambda c + d) = \lambda^2 q(c) + 2\lambda b(c, d) + q(d)$$

has at least one real root. If $q(\lambda c + d) = 0$ then span$(\lambda c + d)$ is a one-dimensional linear subspace of E on which q is identically zero. $\qquad\square$

As a consequence of this, properties of regular quadratic spaces are harder to establish than the corresponding properties for inner product spaces, as we shall see.

Suppose that (E, q) is a quadratic space, with basis $(e_1, \ldots, e_d)$. If the associated symmetric bilinear form b is represented by the matrix $B = (b_{ij})$, then

$$q(x) = \sum_{i=1}^{d} \sum_{j=1}^{d} b_{ij} x_i x_j.$$

If (E, q) is a quadratic space, we can use the corresponding bilinear form b and linear mapping $l_b : E \to E'$ to define annihilation operators $\delta_x : \bigwedge E \to \bigwedge E$, for $x \in E$; we simply set $\delta_x = \delta_{l_b(x)}$. Thus

$$\delta_x(x_1 \wedge \cdots \wedge x_k)$$

$$= \frac{1}{k} \sum_{j=1}^{k} (-1)^{j-1} b(x, x_j)(x_1 \wedge \cdots \wedge \check{x}_j \wedge \cdots \wedge x_k)$$

$$= \frac{1}{k!} \sum_{\sigma \in \Sigma_{n+1}} \epsilon(\sigma) b(x, x_{\sigma(1)})(x_{\sigma(2)} \wedge \cdots \wedge x_{\sigma(k+1)}).$$

We therefore have the following version of Theorem 3.4.1.

Theorem 4.1.1 *Suppose that (E, q) is a quadratic space, and that b is the corresponding quadratic form. If $x, y \in E$ then $m_x^2 = \delta_x^2 = 0$, and $m_x \delta_y + \delta_y m_x = b(x, y)I$.*

4.2 Orthogonality

In this section, we shall suppose that (E, q) is a regular quadratic space, with associated bilinear form b. If x and y are in E, we say that x and y are *orthogonal*, and write $x \perp y$, if $b(x, y) = 0$. Since b is symmetric, $x \perp y$ if and only if $y \perp x$. Note that $x \perp y$ if and only if $q(x + y) = q(x) + q(y)$. If A is a subset of E, we define the *orthogonal set* $A^\perp$ by

$$A^\perp = \{x : x \perp a \text{ for all } a \text{ in } A\}.$$

Proposition 4.2.1 *Suppose that A is a subset of a regular quadratic space E, and that F is a linear subspace of E.*
(i) $A^\perp$ is a linear subspace of E.
(ii) If $A \subseteq B$ then $A^\perp \supseteq B^\perp$.
(iii) $A^{\perp\perp} \supseteq A$ and $A^{\perp\perp\perp} = A^\perp$.
(iv) $A^{\perp\perp} = \text{span}(A)$, and $F = F^{\perp\perp}$.
(v) $\dim F + \dim F^\perp = \dim E$.

Proof l_b is a linear isomorphism of E onto E', and $l_b(x)(y) = 0$ if and only if $b(x, y) = 0$, so that all these results follow from the corresponding duality results. $\square$

Proposition 4.2.2 *Suppose that F is a linear subspace of a regular quadratic space (E, q). Then the following are equivalent.*
(i) (F, q) is regular.
(ii) $F \cap F^\perp = \{0\}$.

(iii) $E = F \oplus F^\perp$.

(iv) $(F^\perp, q)$ *is regular.*

Proof An easy exercise. □

4.3 Diagonalization

Theorem 4.3.1 *Suppose that b is a symmetric bilinear form on a real vector space E. There exists a basis $(e_1, \ldots, e_d)$ and non-negative integers p and m, with $p + m = r$, the rank of b, such that if b is represented by the matrix $B = (b_{ij})$ then*

$$b_{ii} = 1 \text{ for } 1 \le i \le p,$$

$$b_{ii} = -1 \text{ for } p + 1 \le i \le p + m, \text{ and}$$

$$b_{ij} = 0 \text{ otherwise.}$$

A basis which satisfies the conclusions of the theorem is called a *standard orthogonal basis*. If (E, q) is a Euclidean space and b is the associated bilinear form then $b_{ii} = 1$ for $1 \le i \le d$; in this case, the basis is called an *orthonormal basis*.

Proof The proof is by induction on d, the dimension of E. The result is vacuously true when $d = 0$. Suppose that the result holds for all spaces of dimension less than d and all symmetric bilinear forms on them. Suppose that (E, q) is a quadratic space of dimension d and that b is the corresponding symmetric bilinear form. We consider three cases. First it may happen that $q(x) = 0$ for all x in E. Then, by Proposition 4.1.1, $b(x, w) = 0$ for all x and w, and any basis will do: $p = m = r = 0$. Secondly, there exists x with $q(x) > 0$. In this case we set $e_1 = x/\sqrt{q(x)}$. Thirdly, $q(x) \le 0$ for all x, but there exists w with $q(w) < 0$. In this case we set $e_1 = w/\sqrt{-q(w)}$, so that $q(e_1) = -1$.

We now argue in the same way for the last two cases. Let $F = \{e_1\}^\perp$. Then $\dim F = d - 1$, and $E = \operatorname{span}(e_1) \oplus F$. The restriction of b to F is again a symmetric bilinear form: by the inductive hypothesis there is a standard orthogonal basis $(e_2, \ldots, e_d)$ for F. If $j > 1$ then $e_j \in F$, and so $b(e_1, e_j) = b(e_j, e_1) = f(e_j) = 0$. It therefore follows that $(e_1, \ldots, e_d)$ is a standard orthogonal basis for E. It is then clear that $p + m$ is the rank of b. □

If (e_i) is a standard orthogonal basis, and if $x = \sum_{i=1}^{d} x_i e_i$ and

$y = \sum_{i=1}^{d} y_i e_i$ then

$$b(x,y) = \sum_{i=1}^{p} x_i y_i - \sum_{i=p+1}^{p+m} x_i y_i, \quad \text{and} \quad q(x) = \sum_{i=1}^{p} x_i^2 - \sum_{i=p+1}^{p+m} x_i^2.$$

Theorem 4.3.1 depends on the fact that every positive real number has a square root, and every negative real number does not. For more general fields, there exists a basis $(e_1, \ldots, e_d)$ of E, and scalars $(\lambda_1, \ldots, \lambda_d)$ for which $q(x) = \sum_{j=1}^{d} \lambda_j x_j^2$ for all $x = \sum_{j=1}^{d} x_j e_j \in E$.

Theorem 4.3.2 (Sylvester's law of inertia) *Suppose that* $(e_1, \ldots, e_d)$ *and* $(f_i, \ldots, f_d)$ *are standard orthogonal bases for a quadratic space* (E, q), *with corresponding parameters* (p, m) *and* (p', m') *respectively. Then* $p = p'$ *and* $m = m'$.

Proof Let $U = \text{span}\{e_1, \ldots, e_p\}$ and $W = \text{span}\{f_{p'+1}, \ldots, f_d\}$. Then the restriction of q to U is positive definite, while its restriction to W is negative semi-definite. Thus $U \cap W = \{0\}$ and so

$$p + (d - p') = \dim U + \dim W \le d = \dim V.$$

Thus $p \le p'$. Similarly, $p' \le p$, and so $p = p'$. Since $p + m = p' + m' = r$, it follows that $m = m'$. $\square$

We call (p, m) the *signature* $s(q)$ of q. Conventions concerning the signature vary. Many authors write (p, q) for our (p, m) (using some other letter, such as Q, for the quadratic form), so that

$$Q(x) = \sum_{i=1}^{p} x_i^2 - \sum_{i=p+1}^{p+q} x_i^2.$$

Unfortunately, many other authors interchange p and q, and rearrange the basis, so that we get

$$Q(x) = - \sum_{i=1}^{p} x_i^2 + \sum_{i=p+1}^{p+q} x_i^2.$$

We use the letters p and m to stand for 'plus' and 'minus'.

If (E_1, q_1) and (E_2, q_2) are quadratic spaces, then the direct sum $E_1 \oplus E_2$ becomes a quadratic space when we define $q(x_1 \oplus x_2) = q_1(x_1) + q_2(x_2)$. If $(E, q) = (E_1, q_1) \oplus (E_2, q_2)$ then clearly the signature (p, m) of (E, q) is $(p_1 + p_2, m_1 + m_2)$, where (p_1, m_1) is the signature of (E_1, q_1) and (p_2, m_2) is the signature of (E_2, q_2). Thus we have the following corollary.

Corollary 4.3.1 *If F and G are regular subspaces of (E,q) with the same signature, then $F^{\perp}$ and $G^{\perp}$ have the same signature.*

Clearly q is positive definite if and only if $p = d$ and $m = 0$, and is negative definite if and only if $p = 0$ and $m = d$.

We define the *Witt index* to be $w = \min(p,m)$. If $w > 0$ we call (E,q) a *Minkowski* space.

In relativity theory a four-dimensional real vector space with one 'time-like' dimension and three 'space-like' dimensions plays a fundamental rôle. Here, conventions vary: many authors consider quadratic forms with $p = 1$ and $m = 3$, but many others consider forms with $p = 3$ and $m = 1$. We call a regular quadratic space with $m = 1$ a *Lorentz* space.

One other special case occurs, when $p = m$ and $d = p + m = 2p = 2m$. In this case we say that (E,q) is a *hyperbolic* space.

Proposition 4.3.1 *Suppose that (E,q) is a hyperbolic space of dimension $d = 2p$. Then there exists a basis $(f_1,\ldots,f_d)$ such that*

$$b(f_{2i}, f_{2i-1}) = b(f_{2i-1}, f_{2i}) = 1 \quad for \ \ 1 \le i \le p,$$

and $b_{ij} = 0$ otherwise.

Proof Let $(e_1,\ldots,e_d)$ be a standard orthogonal basis for E. Let

$$f_{2i-1} = (1/\sqrt{2})(e_i + e_{p+i}) \quad \text{and} \quad f_{2i} = (1/\sqrt{2})(e_i - e_{p+i})$$

for $1 \le i \le p$. Then

$$b(f_{2i-1}, f_j) = b(f_{2i}, f_j) = 0 \ \ \text{if} \ \ j < 2i-1 \ \text{or} \ j > 2i,$$
$$b(f_{2i-1}, f_{2i-1}) = b(f_{2i}, f_{2i}) = \tfrac{1}{2}(q(e_i) + q(e_{p+1})) = 0,$$
$$b(f_{2i}, f_{2i-1}) = b(f_{2i-1}, f_{2i}) = \tfrac{1}{2}(q(e_i) - q(e_{p+1})) = 1.$$

$\square$

Such a basis is called a *hyperbolic basis*. If $x = \sum_{i=1}^{d} x_i f_i$ and $y = \sum_{i=1}^{d} y_i f_i$ then

$$b(x,y) = (x_1 y_2 + x_2 y_1) + \cdots + (x_{d-1} y_d + x_d y_{d-1}),$$
$$q(x) = 2(x_1 x_2 + x_3 x_4 + \cdots + x_{d-1} x_d).$$

We therefore have the following special cases:

$p = d, m = 0$	Euclidean space
$\min(p, m) > 0$	Minkowski space
$p = d - 1, m = 1$	Lorentz space
$p = m, d = 2p$	Hyperbolic space

It is convenient to have standard examples of the various regular quadratic spaces that can occur. Suppose that p and m are non-negative integers with $p + m = d$. Let $\mathbf{R}_{p,m}$ denote $\mathbf{R}^d$ equipped with the quadratic form

$$q(x) = \sum_{i=1}^{p} x_i^2 - \sum_{i=p+1}^{p+m} x_i^2.$$

$\mathbf{R}_{p,m}$ is called the *standard regular quadratic space* with dimension d and signature (p, m). We write $\mathbf{R}_d$ for $\mathbf{R}_{d,0}$, and call it *standard Euclidean space*; similarly, we use terms such as *standard Minkowski space* and *standard Lorentz space*. Finally, we write H_{2p} for $\mathbf{R}^{2p}$ with the hyperbolic quadratic form

$$q(x) = 2(x_1 x_2 + \cdots + x_{2p-1} x_{2p});$$

this is *standard hyperbolic space*.

Exercises

1. Suppose that (E, q) is a quadratic space. Let

$$N = \{x \in E : b(x, y) = 0 \text{ for all } y \in E\},$$

   and let F be any subspace of E complementary to N. Show that the restriction of q to F is non-singular.

2. Suppose that (E, q) is a regular quadratic space, that H is a hyperbolic subspace of E of maximal dimension and that G is any subspace of E complementary to H. Show that the restriction of q to G is either positive definite or negative definite.

3. Suppose that $\mathbf{Z}_p$ is the field with p elements, where $p = 4k - 1$ is a prime number. Prove a result corresponding to Theorem 4.3.1 for a quadratic form on a finite-dimensional vector space over $\mathbf{Z}_p$.

4.4 Adjoint mappings

Suppose that (E, q) is a regular quadratic space, with associated bilinear form b. Then l_b is an injective linear mapping of E into E', and $\dim E = \dim E'$, and so l_b is an isomorphism of E onto E'. This enables us to define the adjoint of a linear mapping from E into a quadratic space.

Theorem 4.4.1 *Suppose that T is a linear mapping from a regular quadratic space (E_1, q_1) into a quadratic space (E_2, q_2). Then there exists a unique linear mapping T^a from E_2 to E_1 such that*

$$b_2(T(x), y) = b_1(x, T^a(y)) \ \text{ for } \ x \in E_1, y \in E_2,$$

where b_1 and b_2 are the associated bilinear forms.

Proof l_{b_1} is an isomorphism of E_1 onto E_1'. Let $T^a = l_{b_1}^{-1} T' l_{b_2}$. Then it is easy to verify that $b_2(T(x), y) = b_1(x, T^a(y))$ for $x \in E_1$, $y \in E_2$. If S is another linear mapping with this property then

$$b_1(x, T^a(y)) = b_1(x, S(y)) \ \text{ for all } \ x \in E_1, y \in E_2.$$

Since b_1 is non-singular, $T^a(y) = S(y)$ for all $y \in E_2$, and so $T^a = S$. Thus T^a is unique.   □

T^a is called the *adjoint* of T. The mapping $T \rightarrow T^a$ is a linear mapping from $L(E_1, E_2)$ into $L(E_2, E_1)$. If (E_2, q_2) is also regular then $T^{aa} = (T^a)^a = T$, and so the mapping $T \rightarrow T^a$ is an isomorphism of $L(E_1, E_2)$ onto $L(E_2, E_1)$. Note also that if $S \in L(E_2, E_3)$ and $T \in L(E_1, E_2)$, where (E_1, q_1), (E_2, q_2) and (E_3, q_3) are regular quadratic spaces, then $(ST)^a = T^a S^a$.

Suppose that (E, q) is a regular quadratic space with associated bilinear form b, and with standard orthogonal basis $(e_1, \ldots, e_d)$, and suppose that $(\phi_1, \ldots, \phi_d)$ is the dual basis. Then $l_b(e_i) = q(e_i)\phi_i$, so that care is needed with signs. Thus we have the following proposition.

Proposition 4.4.1 *Suppose that (E_1, q_1) and (E_2, q_2) are regular quadratic spaces with standard orthogonal bases $(e_1, \ldots, e_d)$ and $(f_1, \ldots, f_g)$ respectively. Suppose that $T \in L(E_1, E_2)$ and that T is represented by the matrix (t_{ij}) with respect to these bases. Then T^a is represented by the matrix (t_{ji}^a), where*

$$t_{ji}^a = q_1(e_j) q_2(f_i) t_{ij}.$$

Corollary 4.4.1 *If $T \in L(E)$ then $\det T^a = \det T$.*

The matrix (t_{ij}^a) is called the *adjoint matrix* of (t_{ij}). It is the same as the transposed matrix if and only if (E_1, q_1) and (E_2, q_2) are both Euclidean, or q_1 and q_2 are both negative definite.

Exercise

1. We can use the adjoint to consider the relation between creation and annihilation operators. Let q be *any* positive definite quadratic form on E, with corresponding bilinear form b and linear mapping $l_b : E \to E'$. If $x \in E$, let $\delta_x = \delta_{l_b(x)}$. Let $(e_1, \ldots, e_d)$ be a standard orthogonal basis for E. We can define a quadratic form $\tilde{q}$ on $\bigwedge E$ by taking vectors $e_{j_1} \wedge \ldots \wedge e_{j_k}$, with $1 \le j_1 < \cdots < j_k \le d$, as an orthonormal basis.

   Show that $m_x^a = \delta_x$. Note that this gives another proof that $\delta_x^2 = 0$.

4.5 Isotropy

In this section we shall suppose that (E, q) is a regular quadratic space of dimension d and signature (p, m), with associated bilinear form b. An element x of (E, q) is *isotropic* if $q(x) = 0$ and is *anisotropic* otherwise. We write $\mathrm{Iso}(E, q)$ for the set of isotropic elements of E and $\mathrm{An}(E, q)$ for the set of anisotropic elements. $\mathrm{Iso}(E, q)$ is a linear subspace of E only in the trivial cases when $q = 0$ or when (E, q) is positive definite or negative definite; $\mathrm{An}(E, q)$ is never a linear subspace of E, since 0 is isotropic. A linear subspace U of E is *totally isotropic* if it is contained in $\mathrm{Iso}(E, q)$.

Proposition 4.5.1 *A linear subspace U of a regular quadratic space (E, q) is totally isotropic if and only if $U \subseteq U^\perp$.*

Proof The condition is certainly sufficient. If U is totally isotropic then it follows from the polarization formula that the symmetric bilinear form b vanishes identically on $U \times U$, and so $U \subseteq U^\perp$. $\qquad\square$

Proposition 4.5.2 *If U is a totally isotropic subspace of a regular quadratic space E then $\dim U \le w = \min(p, m)$.*

Proof Let $(e_1, \ldots, e_d)$ be a standard orthogonal basis for E, and let $P = \mathrm{span}(e_1, \ldots, e_p)$. Since q is positive definite on P, $P \cap U = \{0\}$. Thus $p + \dim U = \dim P + \dim U \le d$, and so $\dim U \le d - p = m$. Similarly $\dim U \le p$. $\qquad\square$

Recall that $w = \min(p, m)$ is the Witt index of q. If $(e_1, \ldots, e_d)$ is a standard orthogonal basis for (E, q),and if

$$U = \operatorname{span}(e_1 + e_{p+1}, \ldots, e_w + e_{p+w}), \quad W = \operatorname{span}(e_1 - e_{p+1}, \ldots, e_w - e_{p+w})$$

then U and W are both totally isotropic subspaces of E. Thus the Witt index is the best possible upper bound in Proposition 4.5.2. Note also that $U + W = U \oplus W$ is a hyperbolic subspace of E. This is quite typical, as the next result shows.

Proposition 4.5.3 *Suppose that U is a totally isotropic subspace of a regular quadratic space (E, q). Then there exists a totally isotropic space W, with $\dim W = \dim U$, such that $U + W = U \oplus W$ and $U \oplus W$ is hyperbolic.*

Proof Let $(u_1, \ldots, u_r)$ be a basis for U. We show by induction that there exist vectors $(w_1, \ldots, w_r)$ such that

$$b(u_j, w_j) = 1 \text{ for } 1 \le j \le r,$$
$$b(u_i, w_j) = 0 \text{ for } 1 \le i, j \le r \text{ and } i \ne j,$$
$$\text{and } b(w_i, w_j) = 0 \text{ for } 1 \le i, j \le r.$$

Suppose that we have found $(w_1, \ldots, w_s)$ satisfying these conditions, where $0 \le s \le r$. Let $W_s = \operatorname{span}(w_1, \ldots, w_s)$. Suppose that $u = \sum_{j=1}^{s} \alpha_j w_j \in U \cap W_s$. Then $\alpha_j = b(u_j, u) = 0$ for $1 \le j \le s$, and so $u = 0$. Thus $U \cap W_s = 0$, so that $U + W_s = U \oplus W_s$. Further, if $w = \sum_{j=1}^{s} \alpha_j w_j = 0$, then $b(u_j, w) = \alpha_j = 0$, so that the vectors $w_1, \ldots, w_s$ are linearly independent, and $(w_1, \ldots, w_s)$ is a basis for W_s.

Now let $U_{s+1} = \operatorname{span}(u_1, \ldots, \breve{u}_{s+1}, \ldots, u_r)$, where the term u_{s+1} is omitted: U_{s+1} is an $r - 1$ dimensional subspace of E. Then $U_{s+1} \oplus W_s$ is a proper subspace of $U \oplus W_s$, and so $(U \oplus W_s)^{\perp}$ is a proper subspace of $(U_{s+1} \oplus W_s)^{\perp}$. Pick $v \in (U_{s+1} \oplus W_s)^{\perp} \setminus (U \oplus W_s)^{\perp}$. Then $b(v, u_j) = 0$ for $j \ne s + 1$, and $b(v, w) = 0$ for $w \in W_s$. Since $v \notin (U \oplus W_s)^{\perp}$, $b(v, u_{s+1}) \ne 0$. Set $v' = v/b(v, u_{s+1})$, so that $b(v', u_{s+1}) = 1$. We now take $w_{s+1} = v' - b(v', v')u_{s+1}/2$. Then

$$b(w_{s+1}, w_{s+1}) = b(v', v') - b(v', v')b(v', u_{s+1}) + b(v', v')^2 b(u_{s+1}, u_{s+1})/4$$
$$= 0,$$

and, by the construction, all the other conditions are also satisfied. Thus the induction is established.

We now take $W = W_r$. Then $U + W = U \oplus W$, W is totally isotropic, and $(u_1, w_1, \ldots, u_r, w_r)$ is a hyperbolic basis for $U \oplus W$. $\square$

4.6 Isometries and the orthogonal group

We now consider structure-preserving linear mappings. Suppose that (E_1, q_1) and (E_2, q_2) are quadratic spaces, with associated symmetric bilinear forms b_1 and b_2. A linear mapping T from E_1 to E_2 is an *isometry* if

(i) T is injective, and

(ii) $q_2(T(x)) = q_1(x)$ for all x in E_1.

By the polarization formula, (ii) is equivalent to

(iii) $b_2(T(x), T(y)) = b_1(x, y)$ for all x, y in E_1.

Note that an isometry has to be injective. If (E, q) is regular, this is a consequence of (ii) (or (iii)), but if q is singular then this is not the case.

As an easy example, which we shall use later, suppose that (E_1, q_1) has signature (p_1, m_1), that (E_2, q_2) has signature (p_2, m_2), and that $p_1 \leq p_2$, $m_1 \leq m_2$ and $n_1 = d_1 - p_1 - m_1 \leq n_2 = d_2 - p_2 - m_2$. Let $(e_1, \ldots, e_{d_1})$ be a standard orthogonal basis for (E_1, q_1), and let $(f_1, \ldots, f_{d_2})$ be a standard orthogonal basis for (E_2, q_2). If $x = \sum_{j=1}^{d_1} \alpha_j e_j$, let

$$T(x) = \sum_{j=1}^{p_1} \alpha_j f_j + \sum_{j=1}^{m_1} \alpha_{p_1+j} f_{p_2+j} + \sum_{j=1}^{n_1} \alpha_{p_1+m_1+j} f_{p_2+m_2+j}.$$

Then T is an isometry.

The composition of two isometries is an isometry.

Suppose now that (E, q) is a quadratic space. Then an isometry T of (E, q) into itself is called an *orthogonal mapping*. It is necessarily surjective, and its inverse T^{-1} is also an isometry. The orthogonal mappings of E onto itself form a group, the *orthogonal group* $O(E, q)$.

When $(E, q) = \mathbf{R}_{p,m}$, we write $O(p, m)$ for $O(E, q)$. Since $O(E, q) = O(E, -q)$, the groups $O(p, m)$ and $O(m, p)$ are naturally isomorphic. When $(E, q) = \mathbf{R}_d$, we call it the *Euclidean group* $O(d)$; otherwise it is called the *Minkowski group*. Similarly, when $m = 1$ it is called the *Lorentz group*, and when $p = m$ it is called the *hyperbolic group*.

Proposition 4.6.1 *Suppose that $T \in L(E)$, where (E, q) is a regular quadratic space. Then the following are equivalent:*

(i) $T \in O(E, q)$;

(ii) $T^a \in O(E, q)$;

(iii) $T^a T = T T^a = I$.

Proof If $T \in O(E, q)$ then

$$b(x, y) = b(T(x), T(y)) = b(x, T^a T(y)) \text{ for all } x, y \in E,$$

so that $T^a T = I$. Thus $T^{-1} = T^a$, and so $TT^a = I$. Thus (i) implies (iii). Conversely if (iii) holds then

$$b(x, y) = b(x, T^a T(y)) = b(T(x), T(y)) \text{ for all } x, y \in E,$$

so that (i) and (iii) are equivalent. Applying this equivalence to T^a, and using the fact that $T = T^{aa}$, we also obtain the equivalence of (ii) and (iii). □

Corollary 4.6.1 (Orthogonality relations) *Suppose that (E, q) is a regular quadratic space with standard orthogonal basis $(e_1, \ldots, e_d)$. Suppose that $T \in L(E)$ and that T is represented by the matrix (t_{ij}) with respect to this basis. Then the following are equivalent:*

(i) $T \in O(E, q)$;
(ii) $\sum_{i=1}^{d} t_{ij} t_{ik} = \delta_{jk} q(e_j)$ for $1 \leq j \leq d, 1 \leq k \leq d$;
(iii) $\sum_{i=1}^{d} t_{ji} t_{ki} = \delta_{jk} q(e_j)$ for $1 \leq j \leq d, 1 \leq k \leq d$, where $\delta_{jk} = 1$ if $j = k$ and $\delta_{jk} = 0$ otherwise.

Corollary 4.6.2 *If $T \in O(E, q)$ then $\det T = \pm 1$.*

The subgroup $SO(E, q) = \{T \in O(E, q) : \det T = 1\}$ is called the *special orthogonal group*. Let $T(\sum_{j=1}^{d} \alpha_j e_j) = \sum_{j=1}^{d-1} \alpha_j e_j - \alpha_d e_d$. Then $T \in O(E, q)$ and $\det T = -1$, and so $SO(E, q)$ is a proper subgroup of $O(E, q)$. Since $SO(E, q)$ is the kernel of the homomorphism $\det$, which maps $O(E, q)$ onto $\{1, -1\}$, it has index 2 in $O(E, q)$, and we have a short exact sequence

$$1 \longrightarrow SO(E, q) \overset{\subseteq}{\rightarrow} O(E, q) \overset{\det}{\longrightarrow} D_2 \longrightarrow 1$$

Thus if $T \in O(E, q) \setminus SO(E, q)$ then

$$O(E, q) = SO(E, q) \cup T.SO(E, q) = SO(E, q) \cup SO(E, q).T.$$

Proposition 4.6.2 *Suppose that (E, q) is a regular quadratic space, that $T \in O(E, q)$ and that F is a regular subspace of E. Then T is an isometry of F onto $T(F)$ and is an isometry of $F^{\perp}$ onto $(T(F))^{\perp}$.*

Proof Clearly T is is an isometry of F onto $T(F)$ and is an isometry of $F^{\perp}$ onto $T(F^{\perp})$. We must show that $T(F^{\perp}) = (T(F))^{\perp}$. If $x \in F$ and $y \in F^{\perp}$ then $b(T(x), T(y)) = b(x, y) = 0$; since this holds for all $x \in F$, $T(y) \in (T(F))^{\perp}$, and $T(F^{\perp}) \subseteq (T(F))^{\perp}$. Conversely if $z \in (T(F))^{\perp}$

then $z = T(w)$ for some $w \in E$; if $x \in F$ then $b(x,w) = b(T(x),z) = 0$; since this holds for all $x \in F$, $w \in F^{\perp}$, so that $z \in T(F^{\perp})$, and $(T(F))^{\perp} \subseteq T(F^{\perp})$. □

Theorem 4.6.1 (Witt's extension theorem) *Suppose that F is a linear subspace of a regular quadratic space (E, q), and that $T : F \to E$ is an isometry. Then there exists $S \in O(E, q)$ which extends T; that is, $S(x) = T(x)$ for all $x \in F$.*

Proof First suppose that (F, q) is regular, so that $E = F \oplus F^{\perp}$. Then $T(F)$ is regular, and so $E = T(F) \oplus (T(F))^{\perp}$. Since F and $T(F)$ have the same signature, $F^{\perp}$ and $(T(F))^{\perp}$ also have the same signature. Thus, by the example above, there is an isometry R of $F^{\perp}$ onto $(T(F))^{\perp}$. If $x = y + z \in F \oplus F^{\perp}$, let $S(x) = T(x) + R(y)$. Then S is an isometric extension of T.

Next suppose that F is not regular, and that it has signature (p, m). Let $n = d - (p + m)$, where $d = \dim F$. Let $(e_1, \ldots, e_d)$ be a standard orthogonal basis for F. Then $F = G \oplus H$, where $G = \operatorname{span}(e_1, \ldots, e_{p+m})$ is regular, and $H = \operatorname{span}(e_{p+m+1}, \ldots, e_d)$ is totally isotropic. H is a subspace of the regular space $G^{\perp}$, and so by Proposition 4.5.3 there exists a totally isotropic subspace K of $G^{\perp}$ such that $H + K = H \oplus K$ is hyperbolic. Similarly, there exists a totally isotropic subspace L of $(T(G))^{\perp}$ such that $T(H) + L = T(H) \oplus L$ is hyperbolic. Suppose that $(h_1, \ldots, h_n)$ is a basis for H. Then there exists a basis $(k_1, \ldots, k_n)$ for K such that $(h_1, k_1, \ldots, h_n, k_n)$ is a hyperbolic basis for $H \oplus K$. $(T(h_1), \ldots, T(h_n))$ is a basis for $T(H)$, and so there exists a basis $(l_1, \ldots, l_n)$ for L such that $(T(h_1), l_1, \ldots, T(h_n), l_n)$ is a hyperbolic basis for $T(H) \oplus L$. If

$$x = y + \sum_{j=1}^{n} \alpha_j h_j + \sum_{j=1}^{n} \beta_j k_j \in G \oplus H \oplus K = F \oplus K,$$

let $\tilde{T}(x) = T(y) + \sum_{j=1}^{n} \alpha_j T(h_j) + \sum_{j=1}^{n} \beta_j l_j \in T(F) \oplus L.$

Then $\tilde{T}$ is an isometry of $F \oplus K$ onto $T(F) \oplus L$. Since $F \oplus K = G \oplus (H \oplus K)$ is regular, $\tilde{T}$ can be extended to an element S of $O(E, q)$. □

4.7 The case $d = 2$

We now consider $O(E, q)$ and $SO(E, q)$ when E is a two-dimensional regular quadratic space.

The groups $SO(2)$ and $O(2)$

We begin with $SO(2)$. If $T \in SO(2)$ then $q(T(e_1)) = 1$, so that $T(e_1) = (\cos\theta, \sin\theta)$, for some unique $\theta \in [0, 2\pi)$. Since $q(T(e_2)) = 1$, $b(T(e_1), T(e_2)) = 0$ and $\det T = 1$, it follows that $T(e_2) = (-\sin\theta, \cos\theta)$. T is a rotation R_θ of $\mathbf{R}^2$ through an angle θ. Simple calculations show that $R_\theta R_\phi = R_\psi$, where $\psi = \theta + \phi \pmod{2\pi}$, so that the mapping $\theta \to R_\theta$ is an isomorphism of the circle group $\mathbf{T}$ onto $SO(2)$. In particular, this shows that $SO(2)$ is a compact path-connected subset of $L(\mathbf{R}^2)$. Note that $R_\pi = -I$.

Next, suppose that $T \in O(2) \setminus SO(2)$. As before, $T(e_1) = (\cos\theta, \sin\theta)$, for some unique $\theta \in [0, 2\pi)$, but since $\det T = -1$ it follows that $T(e_2) = (\sin\theta, -\cos\theta)$. T has eigenvalues 1 and -1, with corresponding eigenvectors $(\cos\theta/2, \sin\theta/2)$ and $(-\sin\theta/2, \cos\theta/2)$, so that T is an orthogonal reflection S_θ of the plane, leaving the line $\{(\lambda\cos\theta/2, \lambda\sin\theta/2) : \lambda \in \mathbf{R}\}$ fixed.

S_0 is the reflection given by the matrix

$$U = \begin{bmatrix} 1 & 0 \\ 0 & -1 \end{bmatrix},$$

and $O(2) = SO(2) \cup S_0(SO(2))$. Since $R_\theta S_0 R_\theta = S_0$, it follows that $O(2)$ is isomorphic to the full dihedral group D.

The groups $SO(1,1)$ and $O(1,1)$

The quadratic space $\mathbf{R}_{1,1}$ is the simplest example of a hyperbolic space. Here things are more complicated, essentially because a real hyperbola has two connected components. We can consider either the standard orthogonal basis (e_1, e_2) or a standard hyperbolic basis (f_1, f_2).

We begin with the former. Suppose that $T \in SO(1,1)$. Since $q(T(e_1)) = 1$, either

$$T(e_1) = (\cosh t, \sinh t) \quad \text{or} \quad T(e_1) = (-\cosh t, -\sinh t)$$

for some $t \in \mathbf{R}$. We begin with the former case. Since $q(T(e_2)) = -1$, and $\det T = 1$, $T(e_2) = (\sinh t, \cosh t)$. We denote this operator by T_t. T_t is a linear operator on $\mathbf{R}^2$. For $\lambda > 0$, it maps each branch

$$C_\lambda^+ = \{(x, y) : x^2 - y^2 = \lambda, x \geq 0\} \text{ and}$$
$$C_\lambda^- = \{(x, y) : x^2 - y^2 = \lambda, x \leq 0\}$$

of the hyperbola $C_\lambda = \{(x, y) : x^2 - y^2 = \lambda\}$ into itself, and for $\lambda < 0$

maps each branch

$$C_\lambda^+ = \{(x,y) : x^2 - y^2 = \lambda, y \geq 0\} \text{ and}$$
$$C_\lambda^- = \{(x,y) : x^2 - y^2 = \lambda, y \leq 0\}$$

of the hyperbola $C_\lambda = \{(x,y) : x^2 - y^2 = \lambda\}$ into itself. Again, $T_s T_t = T_{s+t}$, so that the mapping $t \to T_t$ is an isomorphism of $(\mathbf{R}, +)$ onto a path-connected subgroup $SO_c(1,1)$ of $SO(1,1)$. Next suppose that $T \in SO(1,1)$ and that $T(e_1) = (-\cosh t, -\sinh t)$. Since $\det T = 1$ it follows that $T(e_2) = (-\sinh t, -\cosh t)$, and that $T = -T_t$; T maps C_λ^+ onto C_λ^-, and C_λ^- onto C_λ^+. Thus $SO(1,1) = SO_c(1,1) \cup -SO_c(1,1)$, and $SO(1,1)$ has two connected components, $SO_c(1,1)$ and $-SO_c(1,1)$. Let $\pi(t) = T_{\log t}$ for $t > 0$ and let $\pi(t) = -T_{\log |t|}$ for $t < 0$. Then π is an isomorphism of the multiplicative group $(\mathbf{R}^*, \times)$ onto $SO(1,1)$.

We now consider $O(1,1)$. The mapping S_0, which maps (x_1, x_2) to $(x_1, -x_2)$ is in $O(1,1)$ and has determinant -1, so that

$$O(1,1) = SO(1,1) \cup S_0.SO(1,-1).$$

Consequently $O(1,1)$ has four connected components, each homeomorphic to $\mathbf{R}$; $SO_c(1,1)$ is the connected component to which the identity belongs, and is a normal subgroup of $O(1,1)$. Thus we have the following short exact sequence:

$$1 \longrightarrow (\mathbf{R}, +) \to O(1,1) \to D_2 \times D_2 \longrightarrow 1.$$

Easy calculations show that $S_0 T_t S_0 = T_{-t}$, so that $(S_0 T_t)^2 = I$, and similarly $(-S_0 T_t)^2 = I$, so that every element of $O(1,1) \setminus SO(1,1)$ is a reflection. The line $\{(\lambda \cosh(t/2), -\lambda \sinh(t/2)) : \lambda \in \mathbf{R}\}$ is fixed by $S_0 T_t$, and $S_0 T_t(\sinh(t/2), -\cosh(t/2)) = -(\sinh(t/2), -\cosh(t/2))$. Since $T_t = S_0(S_0 T_t)$ and $-T_t = S_0(S_0(-T_t))$, every element of $SO(1,1)$ is the product of two reflections in $O(1,1)$.

Next, we consider the hyperbolic basis (f_1, f_2). Then $\text{Iso}(\mathbf{R}_{1,1}) = \text{span}(f_1) \cup \text{span}(f_2)$. Suppose that $T \in O(1,1)$. Since $q(f_1) = 0$, it follows that $q(T(f_1)) = 0$, so that either $T(f_1) = af_1$ or $T(f_1) = af_2$, where $a \in \mathbf{R}^*$. In the former case, $T(f_2) = f_2/a$, since $q(T(f_2)) = q(f_2) = 0$ and $b(T(f_1), T(f_2)) = b(f_1, f_2) = 1$. Thus T is represented by the matrix

$$\begin{bmatrix} a & 0 \\ 0 & 1/a \end{bmatrix},$$

which has determinant 1. Thus in this case $T \in SO(1,1)$. In the latter case, where $T(f_1) = af_2$, it follows that $T(f_2) = f_1/a$; T is represented by the matrix

$$\begin{bmatrix} 0 & 1/a \\ a & 0 \end{bmatrix},$$

which has determinant -1, and so $T \in O(1,1) \setminus SO(1,1)$. Consequently the mapping

$$a \rightarrow \begin{bmatrix} a & 0 \\ 0 & 1/a \end{bmatrix}$$

establishes an isomorphism of $(\mathbf{R}^*, \times)$ onto $SO(1,1)$, and

$$O(1,1) \setminus SO(1,1) = T_1.SO(1,1) = SO(1,1).T_1,$$

where $T_1(x_1 f_1 + x_2 f_2) = x_2 f_1 + x_1 f_2$.

Exercise

1. Let G be the group of homeomorphisms of $\mathbf{R}$ generated by the set of *dilations* $\{D_\lambda : \lambda > 0\}$ (where $D_\lambda(t) = \lambda t$), *negation* N (where $N(t) = -t$) and *inversion* σ (where $\sigma(t) = 1/t$). Show that G is isomorphic to $O(1,1)$.

4.8 The Cartan-Dieudonné theorem

We have seen that every element of $O(2)$ and every element of $O(1,1)$ is either a reflection, or the product of two reflections. Can we extend this result to all regular quadratic spaces? First we must define the reflections that we should consider. Suppose that x is an anisotropic vector in a regular quadratic space (E,q). Then span(x) is regular, and $E = \text{span}(x) \oplus x^{\perp}$. Suppose that $y \in E$, and that $y = \lambda x + z$, with $z \in x^{\perp}$. Let $\rho_x(y) = -\lambda x + z$. Then ρ_x is an injective linear mapping, and

$$q(\rho_x(y)) = \lambda^2 q(x) + q(z) = q(y),$$

so that ρ_x is an isometry. $\rho_x(x) = -x$, and $\rho_x(z) = z$ for $z \in x^{\perp}$. Also $\rho_x^2 = I$ (ρ_x is an *involution*), so that ρ_x is surjective. The mapping ρ_x is called the *simple reflection in the direction* x, with *mirror* $x^{\perp}$. Note that $\det \rho_x = -1$.

Theorem 4.8.1 (Cartan-Dieudonné) *If T is an isometry of a regular quadratic space (E, q), T is the product of at most $\dim E$ simple reflections.*

Proof The proof is by induction on the dimension d of E. The result is clearly true when $d = 1$, and we have seen that it is true when $d = 2$. Suppose that $\dim E = d \geq 3$, and that the result is true for spaces of dimension less than d.

Let us set $S = I - T$, and let $N(S)$ be the null-space of S: $N(S)$ is the set of vectors fixed by T. We consider three cases.

Suppose first that $N(S)$ is not totally isotropic, and that x is an anisotropic element of $N(S)$. We can then write $E = \operatorname{span}(x) \oplus F$, where $F = x^{\perp}$, and T maps F isometrically onto itself. Let S be the restriction of T to F. By the inductive hypothesis, we can write $S = \sigma_{x_r} \ldots \sigma_{x_1}$ where each σ_{x_i} is a simple reflection of F in the direction x_i and $r \leq \dim F = d - 1$. But then it is easy to see that $T = \rho_{x_r} \ldots \rho_{x_1}$, where each ρ_{x_i} is the simple reflection of E in the direction x_i.

Secondly, suppose that there exists an anisotropic vector x such that $y = S(x)$ is anisotropic. Then $b(T(x)+x, T(x)-x) = b(T(x)+x, y) = 0$, so that $T(x) + x \in y^{\perp}$. Thus

$$\rho_y(T(x)) = \rho_y \left(\frac{T(x) + x}{2} \right) + \rho_y \left(\frac{T(x) - x}{2} \right)$$
$$= \frac{T(x) + x}{2} - \frac{T(x) - x}{2} = x,$$

and so by the first case $\rho_y T$ is the product of at most $d-1$ simple reflections. But then $T = \rho_y \rho_y T$ is the product of at most d simple reflections.

(Notice that if q is positive definite (or negative definite) then at least one of these two cases must occur, and the proof is finished.)

Thirdly, suppose that neither of the two conditions is satisfied. We show that this can only happen in very special circumstances. First we show that $S(E)$ is totally isotropic. By hypothesis, $S(x)$ is isotropic if x is anisotropic; it is therefore sufficient to show that if x is a non-zero isotropic vector then $q(S(x)) = 0$. Since $\dim x^{\perp} = d - 1 > d/2$, $x^{\perp}$ is not totally isotropic, and so there exists an anisotropic vector $u \in x^{\perp}$. Now

$$q(x + u) = q(x - u) = q(u) \neq 0,$$

and so, since the second condition is not satisfied,

$$q(S(x + u)) = q(S(x - u)) = q(S(u)) = 0.$$

Now

$$q(S(x+u)) + q(S(x-u)) = 2q(S(x)) + 2q(S(u)) = 2q(S(x)),$$

and so $S(x)$ is isotropic.

Thus $S(E)$ and $N(S)$ are both totally isotropic. Consequently, $\dim S(E) \leq w$ and $\dim N(S) \leq w$, where w is the Witt index of (E, q). But by the rank-nullity formula, $\dim(S(E)) + \dim(N(S)) = d \geq 2w$. Thus $\dim S(E) = \dim N(S) = w = d/2$, so that d is even, and (E, q) is a hyperbolic space. In particular,

$$\dim(N(S)^\perp) = d - \dim N(S) = w = \dim N(S),$$

and so $N(S) = N(S)^\perp$. Now suppose that $w = S(x) \in S(E)$ and that $y \in N(S)$. Then $T(y) = y$, and so

$$b(w, y) = b(T(x), y) - b(x, y) = b(T(x), T(y)) - b(x, y) = 0.$$

Thus $S(E) \subseteq N(S)^\perp = N(S)$, and so $S^2 = 0$.

Now let $(e_1, \ldots, e_w)$ be a basis for $N(S)$, and extend this to a basis $(e_1, \ldots, e_d)$ of E. With respect to this basis, S is given by a matrix of the form

$$\begin{bmatrix} 0 & M \\ 0 & 0 \end{bmatrix},$$

and so T is given by a matrix of the form

$$\begin{bmatrix} I & -M \\ 0 & I \end{bmatrix}.$$

This means that $\det T = 1$. Now let y be any anisotropic vector in E. Then $\rho_y T$ is an isometry, and $\det \rho_y T = -1$. This means that $\rho_y T$ is not in this final case, and so must be in one of the first two cases. Thus $\rho_y T$ is the product of at most $d = 2w$ simple reflections. But since $\det \rho_y T = -1$, $\rho_y T$ must be the product of an odd number of simple reflections, and is therefore the product of at most $d-1$ simple reflections. Thus $T = \rho_y \rho_y T$ is the product of at most d simple reflections. □

Corollary 4.8.1 *(i) If* $\dim E$ *is odd and* $T \in SO(E, q)$, *then there exists a non-zero* x *with* $T(x) = x$.

(ii) If $\dim E$ *is odd and* $T \in O(E, q) \setminus SO(E, q)$, *then there exists a non-zero* x *with* $T(x) = -x$.

(iii) If $\dim E$ *is even and* $T \in O(E, q) \setminus SO(E, q)$, *then there exists a non-zero* x *with* $T(x) = x$.

Proof (i) $T = \rho_{x_r} \ldots \rho_{x_1}$ is the product of an even number r of simple reflections, and $r < \dim E$. Let M_i be the mirror for ρ_{x_i}; then $\dim M_i = \dim E - 1$, and so $\dim(\cap_{i=1}^r M_i) \geq \dim E - r > 0$. If x is a non-zero element of $\cap_{i=1}^r M_i$, $T(x) = x$.

(ii) $-T \in SO(E, q)$, and so the result follows from (i).

(iii) Since $T \in O(E, q) \setminus SO(E, q)$, T is the product of an odd number s of simple reflections, and $s < \dim(E)$. Now argue as in (i). □

As we have observed, if $T \in SO(E, q)$ then T is the product of an even number of simple reflections. We group these in pairs: we call the product of two simple reflections a *simple rotation*. Thus every element of $SO(E, q)$ is the product of at most $d/2$ simple rotations. Let us consider simple rotations in more detail. Suppose that $R = \rho_x \rho_y$ is a simple rotation. Let $F = \operatorname{span}(x, y)$. Then F is a regular subspace of E, and $E = F \oplus F^\perp$. $R(z) = z$ for $z \in F^\perp$. If x and y are linearly dependent, then $R = I$; otherwise the restriction R_F of R to F is a rotation of the two-dimensional space F. Let (e_1, e_2) be an orthogonal basis for F. If (F, q) is positive definite or negative definite then $R_F = R_\theta$, a rotation through an angle θ for some $0 < \theta < 2\pi$. If (F, q) is hyperbolic, then $R_F = T_t$ or $-T_t$, for some $t \in \mathbf{R}$.

The next result is an immediate consequence.

Theorem 4.8.2 *If (E, q) is a Euclidean space, then $SO(E, q)$ is path-connected.*

Exercises

1. Suppose that $S_\theta(z) = e^{i\theta} \bar{z}$ is a reflection of $\mathbf{C}$. Determine the set $\{w \in \mathbf{C} : S_\theta = \rho_w\}$, and determine the mirror of this reflection.

2. Suppose that $T \in O(E, q)$, and that F is a regular subspace of E such that $T(F) \subseteq F$. Show that $T(F^\perp) = F^\perp$.

3. Suppose that $T \in L(E)$, where E is a d-dimensional real vector space.

   (a) Show that there is a non-zero polynomial p of minimal degree such that $p(T) = 0$.

   (b) Show that any real polynomial can be written as a product of real polynomial of degree 1 or 2.

   (c) Show that there is a linear subspace F of E of dimension 1 or 2 such that $T(F) \subseteq F$.

   (d) Suppose that (E, q) is a Euclidean space, and that $T \in SO(E, q)$. Show that there exists an orthogonal basis for E with respect to which

T is represented by a matrix of the form

$$
\begin{bmatrix}
T_1 & \cdots & 0 & 0 \\
\vdots & \ddots & \vdots & \vdots \\
0 & \cdots & T_k & 0 \\
0 & \cdots & 0 & I_r
\end{bmatrix},
$$

where $T_j = \begin{bmatrix} \cos t_j & -\sin t_j \\ \sin t_j & \cos t_j \end{bmatrix}$ for $1 \leq j \leq k$,

I_r is an $r \times r$ unit matrix and $2k + r = d$.

4. Find necessary and sufficient conditions for two simple reflections to commute.

4.9 The groups $SO(3)$ and $SO(4)$

The quaternions can be used to obtain double covers of the groups $SO(3)$ and $SO(4)$. These were described by Cayley, who ascribed the first of them to Hamilton. These double covers will reappear in Chapter 8, when we consider spin groups.

The subspace $(Pu(\mathbf{H}), \Delta)$ of $\mathbf{H}$ is a Euclidean space, isometrically isomorphic to $\mathbf{R}^3$. We can therefore identify $SO(Pu(\mathbf{H}), \Delta)$ with $SO(3)$.

Theorem 4.9.1   *Suppose that $x \in \mathbf{H}^*$. Let*

$$
\rho_x(y) = xyx^{-1} \quad \text{for} \ \ y \in Pu(\mathbf{H}).
$$

Then ρ is a homomorphism of the multiplicative group $\mathbf{H}^*$ onto $SO(3)$, with kernel $\mathbf{R}^*$.

Proof   First we show that $\rho_x(y) \in Pu(\mathbf{H})$:

$$
(\rho_x(y))^2 = xy^2 x^{-1} = -\Delta(y)xx^{-1} = -\Delta(y) \leq 0,
$$

so that this follows from Proposition 2.3.1. Thus $\rho_x \in L(Pu(\mathbf{H}))$. Further,

$$
\Delta(\rho_x(y)) = \Delta(x)\Delta(y)\Delta(x^{-1}) = \Delta(y)\Delta(xx^{-1}) = \Delta(y),
$$

so that $\rho_x \in O(3)$. Thus $\det \rho_x = \pm 1$. But $\mathbf{H}^*$ is connected, and the function det is continuous on $\mathbf{H}^*$, and so $\det \rho_x$ is constant on $\mathbf{H}^*$. Since $\det \rho_1 = \det I = 1$, it follows that $\det \rho_x = 1$ for all $x \in \mathbf{H}^*$, and so ρ is a homomorphism of $\mathbf{H}^*$ into $SO(3)$.

If $x \in \mathbf{H}^*$ then $\rho_x = I$ if and only if $x \in Z(\mathbf{H}^*) = \mathbf{R}^*$, and so $\mathbf{R}^*$ is the kernel of ρ.

It remains to show that $\rho(\mathbf{H}^*) = SO(3)$. Suppose that $x \in Pu(\mathbf{H})$ and that $x \neq 0$. Then $\rho_x(x) = x$, and if $y \in x^\perp$ then

$$\rho_x(y) = xyx^{-1} = -yxx^{-1} = -y.$$

Thus $-\rho_x$ is a simple reflection in the direction x, with mirror $x^\perp$. If $T \in SO(3)$ then T is the product of two simple reflections, and so there exist non-zero x_1, x_2 in $Pu(\mathbf{H})$ such that $T = (-\rho_{x_1})(-\rho_{x_2}) = \rho_{x_1 x_2}$, and $T \in \rho(\mathbf{H}^*)$. □

Thus we have a short exact sequence

$$1 \longrightarrow \mathbf{R}^* \longrightarrow \mathbf{H}^* \longrightarrow SO(3) \longrightarrow 1.$$

Since $\rho_x = \rho_{\lambda x}$, for $\lambda \in \mathbf{R}^*$, $\rho(x) = \rho(x_1) = \rho(-x_1)$, where $x_1 = x/\|x\| \in \mathbf{H}_1$. We therefore also have the following short exact sequence

$$1 \longrightarrow D_2 \longrightarrow \mathbf{H}_1 \longrightarrow SO(3) \longrightarrow 1;$$

$\mathbf{H}_1$ is a double cover of $SO(3)$.

Here is a similar result, concerning $SO(4)$. The quadratic space $(\mathbf{H}, \Delta)$ is a Euclidean space, isometrically isomorphic to $\mathbf{R}^4$, and so we identify $SO(\mathbf{H}, \Delta)$ and $SO(4)$.

Theorem 4.9.2 *Let $\phi_{(x,y)}(z) = xzy^{-1}$, for $(x,y) \in \mathbf{H}_1 \times \mathbf{H}_1$ and $z \in \mathbf{H}$. Then ϕ is a homomorphism of $\mathbf{H}_1 \times \mathbf{H}_1$ onto $SO(4)$, with kernel $\{(1,1), (-1,-1)\}$.*

Proof Since $\|xzy^{-1}\| = \|x\| . \|z\| . \|y^{-1}\| = \|z\|$, $\phi_{(x,y)} \in O(4)$. Thus $\det \phi_{(x,y)} = \pm 1$. The group $\mathbf{H}_1 \times \mathbf{H}_1$ is connected, and so the argument of Theorem 4.9.1 shows that $\phi_{(x,y)} \in SO(4)$. Thus ϕ is a homomorphism of $\mathbf{H}_1 \times \mathbf{H}_1$ into $SO(4)$. If (x,y) is in the kernel of ϕ, then $xy^{-1} = \phi_{(x,y)}(1) = 1$, so that $x = y$. Then $xzx^{-1} = z$ for all $z \in \mathbf{H}$, so that $x \in Z(\mathbf{H}) = \mathbf{R}$. Since $\|x\| = 1$, $x = 1$ or -1. Since $\phi_{(1,1)} = \phi_{(-1,-1)}$, the kernel of ϕ is $\{(1,1), (-1,-1)\}$.

It remains to show that ϕ is surjective. Suppose that $T \in SO(4)$. Let $b = T(1)$. Since $\|b\| = 1$, $b \in \mathbf{H}_1$. Thus $\phi_{(1,b)}T(1) = 1$. Since $\phi_{(1,b)} \circ T \in SO(4)$, it follows that $\phi_{(1,b)}T(Pu(\mathbf{H})) = Pu(\mathbf{H})$, and the restriction of $\phi_{(1,b)} \circ T$ to $Pu(\mathbf{H})$ is in $SO(Pu(\mathbf{H}), \Delta)$. By Theorem 4.9.1, there exist x_1 and x_2 in $\mathbf{H}_1$ so that $\phi_{(1,b)}T = \rho_{x_1}\rho_{x_2} = \phi_{(x_1 x_2, x_1 x_2)}$. Consequently, $T = \phi_{(y,z)}$, where $y = x_1 x_2$ and $z = b^{-1}x_1 x_2$. □

Thus we have a short exact sequence

$$1 \longrightarrow D_2 \longrightarrow \mathbf{H}_1 \times \mathbf{H}_1 \longrightarrow SO(4) \longrightarrow 1;$$

$\mathbf{H}_1 \times \mathbf{H}_1$ is a double cover of $SO(4)$.

4.10 Complex quadratic forms

Our principal concern is with real quadratic spaces. Here, we briefly consider what happens in the complex case. Suppose then that q is a quadratic form on a complex vector space E; that is, there exists a symmetric complex bilinear form b on E such that $q(x) = b(x,x)$, for $x \in E$. The polarization equation holds, and so b is uniquely determined by q. Again, we say that (E,q) is *regular* if b is non-singular. Since every complex number has a square root, following through the proof of Theorem 4.3.1 we obtain the following:

Theorem 4.10.1 *Suppose that b is a symmetric bilinear form on a complex vector space E. There exists a basis $(e_1, \ldots, e_d)$ such that if b is then represented by the matrix $B = (b_{ij})$ then*

$$b_{ii} = 1 \ \ for \ \ 1 \le i \le r,$$

$$b_{ij} = 0 \ \ otherwise,$$

where r is the rank of b.

If $d > 1$ then (E,q) contains isotropic subspaces, even when q is regular; for $\mathrm{span}(e_1 + ie_2)$ is isotropic. In particular, if (E,q) is regular and $d = 2p$ is even, then

$$\left(\frac{e_1 + ie_2}{\sqrt{2}}, \frac{e_1 - ie_2}{\sqrt{2}}, \frac{e_3 + ie_4}{\sqrt{2}}, \frac{e_3 - ie_4}{\sqrt{2}}, \ldots, \frac{e_{2p-1} + ie_{2p}}{\sqrt{2}}, \frac{e_{2p-1} - ie_{2p}}{\sqrt{2}} \right)$$

is a hyperbolic basis for (E,q).

The proofs of Proposition 4.5.3 and the Witt extension theorem carry over to the complex case, as does the proof of the Cartan-Dieudonné theorem.

Let $O(E,q)$ denote the group of isometries of a regular complex quadratic space. If $T \in O(E,q)$ then, by the Cartan-Dieudonné theorem, $\det T = 1$ or -1; again, we denote by $SO(E,q)$ the subgroup

$\{T \in O(E,q) : \det T = 1\}$. The product of two simple reflections is again called a *simple rotation*. Suppose that $R = \rho_x \rho_y$ is a simple rotation, and that x and y are linearly independent. Then $F = \mathrm{span}(x, y)$ is a regular two-dimensional R-invariant subspace, and $R(z) = z$ for $z \in F^\perp$. Let R_F be the restriction of R to F. There are two possibilities.

First, it may happen that R_F has an anistropic eigenvector u, with eigenvalue λ. Since $q(u) = q(T(u)) = q(\lambda u) = \lambda^2 q(u)$, $\lambda = 1$ or -1. Also $u^\perp \cap F$ is a one-dimensional R-invariant space, and so if $v \in u^\perp \cap F$ then $R(v) = v$ or $-v$. Since $\det R = 1$, $R_F = I$ or $-I$. The first possibility can be ruled out, since x and y are linearly independent, so that $R_F = -I$.

The second possibility is that every eigenvector of R_F is isotropic. Let (f_1, f_2) be a hyperbolic basis for F. Then $\mathrm{Iso}(F, q) = \mathrm{span}(f_1) \cup \mathrm{span}(f_2)$, so that f_1 and f_2 must be eigenvectors for R_F. Suppose that $R_F(f_1) = \lambda f_1$ and that $R_F(f_2) = \mu f_2$. Since $\det R_F = 1$, it follows that $\lambda \mu = 1$. It is readily verified that if this is the case, then R_F is an isometry of F. Note that when $\lambda = \mu = \pm 1$ then $R_F = \pm I$. contradicting the fact that every eigenvector of R_F is isotropic. Thus $\lambda \in \mathbf{C}^* \setminus \{1, -1\}$.

Bearing in mind that every element of $L(E)$ has an eigenvector, and that if F is a regular T-invariant subspace of E then $F^\perp$ is a regular T-invariant subspace of E, we have the following.

Theorem 4.10.2 *If (E, q) is a regular complex quadratic space and $T \in O(E, q)$, then $E = F \oplus G \oplus H$ is a direct sum of regular T-invariant subspaces of E, where*

$$ F = \{x \in E : T(x) = x\}, \qquad G\{x \in E : T(x) = -x\}, $$

H has a hyperbolic basis $(h_1, h_2, \ldots, h_{2k-1}, h_{2k})$, and there exist scalars $\lambda_1, \ldots, \lambda_k$ in $\mathbf{C}^* \setminus \{1, -1\}$ such that $T(h_{2i-1}) = \lambda_i h_{2i-1}$ and $T(h_{2i}) = \lambda_i^{-1} h_{2i}$ for $1 \le i \le k$. $T \in SO(E, q)$ if and only if the dimension of G is even.

Exercise

1. In Theorem 4.10.2, let

$$ g_{2j-1} = (h_{2j-1} + h_{2j})/\sqrt{2}, \; g_{2j} = i(h_{2j-1} - h_{2j})/\sqrt{2}, $$

for $1 \le j \le k$. Calculate $T(g_i)$, for $1 \le i \le 2k$.

4.11 Complex inner-product spaces

Besides complex quadratic spaces, we can consider complex inner-product spaces. An inner product on a complex vector space E is a mapping from $E \times E$ into $\mathbf{C}$ which satisfies:

(i)
$$\begin{aligned}
\langle \alpha_1 x_1 + \alpha_2 x_2, y \rangle &= \alpha_1 \langle x_1, y \rangle + \alpha_2 \langle x_2, y \rangle, \\
\langle x, \beta_1 y_1 + \beta_2 y_2 \rangle &= \bar{\beta}_1 \langle x, y_1 \rangle + \bar{\beta}_2 \langle x, y_2 \rangle,
\end{aligned}$$
for all x, x_1, x_2, y, y_1, y_2 in E and all complex $\alpha_1, \alpha_2, \beta_1, \beta_2$;

(ii) $\langle y, x \rangle = \overline{\langle x, y \rangle}$ for all x, y in E, and

(iii) $\langle x, x \rangle > 0$ for all non-zero x in E.

For example, if $E = \mathbf{C}^d$, we define the *usual* inner product by setting $\langle z, w \rangle = \sum_{j=1}^{d} z_j \bar{w}_j$ for $z = (z_j), w = (w_j)$.

A complex vector space E equipped with an inner product is called an *inner-product* space. An inner product is *sesquilinear*: linear in the first variable, and conjugate-linear in the second variable. [A word of warning: this is the convention that most pure mathematicians use. Physicists use the other convention: conjugate-linear in the first variable and linear in the second variable.] The quantity $\|x\| = \langle x, x \rangle^{1/2}$ is then a norm on E, and the function $d : E \times E \to \mathbf{R}$ defined by $d(x, y) = \|x - y\|$ is a metric on E. If E is complete under this metric then E is called a *Hilbert space*. Any finite-dimensional inner-product space is complete; such a space is called a *Hermitian space*.

Arguing as in Theorem 4.10.1, it follows that if $(E, \langle .,. \rangle)$ is an inner-product space then there exists an orthonormal basis for E; a basis $(e_1, \dots, e_d)$ such that

$$\langle e_j, e_j \rangle = 1 \ \text{ for } \ 1 \le j \le d \ \text{ and } \ \langle e_i, e_j \rangle = 0 \ \text{ for } \ i \ne j.$$

In this case,

$$\left\langle \sum_{j=1}^{d} z_j e_j, \sum_{j=1}^{d} w_j e_j \right\rangle = \sum_{j=1}^{d} z_j \bar{w}_j,$$

so that $(E, \langle .,. \rangle)$ is isometrically isomorphic to $\mathbf{C}^d$, with its usual inner product $\langle x, y \rangle = \sum_{j=1}^{d} x_j \bar{y}_j$.

Arguing as in Theorem 4.4.1, if T is a linear mapping from a Hermitian space E to a Hermitian space F then there is a unique linear mapping $T^* : F \to E$ such that

$$\langle T(x), y \rangle = \langle x, T^*(y) \rangle \quad \text{for all} \ \ x \in E, y \in F.$$

T^* is called the *adjoint* of T. (More generally, if E and F are Hilbert spaces and $T : E \to F$ is a bounded linear operator, then T has a unique adjoint $T^* : F \to E$ with the property described above.) If $(e_1, \ldots, e_d)$ is an orthonormal basis for E, if $(f_1, \ldots, f_c)$ is an orthonormal basis for F and if T is represented by the matrix (t_{ij}), then T^* is represented by the matrix (t_{ij}^*), where $t_{ij}^* = \bar{t}_{ji}$.

A linear mapping T from a Hilbert space E to itself is *Hermitian* if $T = T^*$. If E is finite-dimensional and T is represented by a matrix (t_{ij}) with respect to an orthonormal basis then T is Hermitian if and and only if $t_{ij} = \bar{t}_{ji}$ for all i, j; the matrix (t_{ij}) is then said to be Hermitian. If T is Hermitian there exists an orthonormal basis $(e_1, \ldots, e_d)$ and real scalars $(\lambda_1, \ldots, \lambda_d)$ such that $T(e_i) = \lambda_i e_i$ for $1 \leq i \leq d$: $\lambda_1, \ldots, \lambda_d$ are the eigenvalues of T.

The group of linear isometries of $(E, \langle ., . \rangle)$ is denoted by $U(E, \langle ., . \rangle)$, and is called the *unitary group*. We write $U(d)$ for $U(\mathbf{C}^d)$, where $\mathbf{C}^d$ is given its usual inner product. Suppose that $T \in U(E, \langle ., . \rangle)$. If λ is an eigenvalue for T, with eigenspace E_λ, then $|\lambda| = 1$, so that $\lambda = e^{i\theta}$ for some $\theta \in [0, 2\pi)$. Arguing as in Proposition 4.6.2, we see that $T(E_\lambda^\perp) = E_\lambda^\perp$; consequently, there exists an orthonormal basis $(e_1, \ldots, e_d)$ for E and a sequence $(\theta_1, \ldots, \theta_d)$ in $[0, 2\pi)$ such that $T(e_j) = e^{i\theta_j} e_j$ for $1 \leq j \leq d$. A unitary mapping T is in the *special unitary group* $SU(d)$ if $\det T = 1$; in terms of the representation just given, $T \in SU(d)$ if and only if $\theta_1 + \cdots + \theta_d = 0 \pmod{2\pi}$. Both of the groups $U(d)$ and $SU(d)$ are compact and path-connected.

5

Clifford algebras

We have seen that if E is a vector space, then E can be embedded in the tensor algebra $\bigotimes^*(E)$, in the symmetric algebra $\bigotimes^*_s(E)$, and, most pertinently, in the exterior algebra $\bigwedge^*(E)$. The exterior algebra was introduced by Grassmann in 1844. In 1876, Clifford communicated the abstract of a paper [Cli2] to the London Mathematical Society; this paper, which was unfinished, and not published in his lifetime, showed how to modify the definition of the exterior algebra of an n-dimensional real inner-product space, to take the inner product into account. In this paper, he showed that the resulting algebra (which he called a 'geometric algebra') has dimension 2^n, and that it can be decomposed into odd and even parts. He also observed that properties of the algebra depended on the value of $n \pmod 4$.

Clifford's definition extends easily to general quadratic spaces. It also extends to complex quadratic spaces, and indeed to quadratic spaces defined over more general fields. We shall however restrict our attention to real quadratic spaces.

5.1 Clifford algebras

Throughout this chapter, we shall suppose that (E, q) is a d-dimensional real vector space E with quadratic form q, associated bilinear form b and standard orthogonal basis $(e_1, \ldots, e_d)$. Suppose that A is a unital algebra. A *Clifford mapping* j is an injective linear mapping $j : E \to A$ such that

   (i) $1 \notin j(E)$, and

   (ii) $(j(x))^2 = -q(x)1 = -q(x)$, for all $x \in E$.

If, further,

(iii) $j(E)$ generates A,

then A (together with the mapping j) is called a *Clifford algebra* for (E, q).

If j is a Clifford mapping, and $x, y \in E$ then

$$j(x)j(y) + j(y)j(x) = j(x+y)^2 - j(x)^2 - j(y)^2$$
$$= -(q(x+y) + q(x) + q(y))1 = -2b(x,y)1.$$

In particular, if $x \perp y$ then $xy = -yx$.

Conventions vary greatly! Some authors require that $(j(x))^2 = q(x)$.

We can identify $\mathbf{R}$ with span(1): these are the *scalars* in A. We can also identify E with $j(E)$, so that E is a linear subspace of A. Elements of E are then called *vectors* in A.

Let us give three examples. First, suppose that $q = 0$. Let $A = \bigwedge^* E$, and let $j(x) = x$. Since $1 \notin j(E)$, and $x \wedge x = 0 = -q(x)$ for all $x \in E$, j is a Clifford mapping of (E, q) into $\bigwedge^* E$. Since E generates the algebra $\bigwedge^* E$, the exterior algebra $\bigwedge^* E$ is a Clifford algebra for (E, q).

Secondly, Let (E, q) be a one-dimensional inner-product space, with basic element e_1 satisfying $q(e_1) = 1$. Let $A = \mathbf{C}$, the algebra of complex numbers, and let $j(\lambda e_1) = \lambda i$. Then $(j(\lambda e_1))^2 = -\lambda^2 = -q(\lambda e_1)$. Thus j is a Clifford mapping and $\mathbf{C}$ is a Clifford algebra for (E, q). (This example explains the sign-convention that we use.) We can think of $A = \mathbf{C}$ as a subalgebra of $M_2(\mathbf{R})$, with

$$1 = \begin{bmatrix} 1 & 0 \\ 0 & 1 \end{bmatrix} \text{ and } i = J = \begin{bmatrix} 0 & -1 \\ 1 & 0 \end{bmatrix},$$

so that a typical element of A is

$$z = x + iy = xI + yJ = \begin{bmatrix} x & -y \\ y & x \end{bmatrix}.$$

Thirdly, let (E, q) be a one-dimensional quadratic space, with negative-definite form q, and with basic element e_1 satisfying $q(e_1) = -1$. Let $A = \mathbf{R}^2$, with multiplication defined as $(x_1, y_1)(x_2, y_2) = (x_1 x_2, y_1 y_2)$, so that $A = \mathbf{R} \oplus \mathbf{R}$ is the algebra direct sum of two copies of $\mathbf{R}$, with identity element $1 = (1, 1)$. Let $j(\lambda e_1) = (\lambda, -\lambda)$. Then $1 \notin j(E)$, and

$$(j(\lambda e_1))^2 = \lambda^2(1,1) = -q(\lambda e_1).1,$$

so that j is a Clifford mapping into A, and A is a Clifford algebra for (E, q). Again we can think of A as a subalgebra of $M_2(\mathbf{R})$, with

$$1 = \begin{bmatrix} 1 & 0 \\ 0 & 1 \end{bmatrix} \text{ and } j(e_1) = U = \begin{bmatrix} 1 & 0 \\ 0 & -1 \end{bmatrix},$$

so that a typical element of A is

$$xI + yU = \begin{bmatrix} x+y & 0 \\ 0 & x-y \end{bmatrix}.$$

Alternatively, we could take

$$j(\lambda e_1) = \lambda Q = \begin{bmatrix} 0 & \lambda \\ \lambda & 0 \end{bmatrix},$$

so that a typical element of A is

$$xI + yQ = \begin{bmatrix} x & y \\ y & x \end{bmatrix}.$$

We shall make frequent use of the following elementary result.

Theorem 5.1.1 *Suppose that that $a_1, \ldots, a_d$ are elements of a unital algebra A. Then there exists a Clifford mapping $j : (E, q) \to A$ satisfying $j(e_i) = a_i$ for $1 \leq i \leq d$ if and only if*

$$a_i^2 = -q(e_i) \quad \text{for } 1 \leq i \leq d,$$
$$a_i a_j + a_j a_i = 0 \quad \text{for } 1 \leq i < j \leq d,$$
$$\text{and } 1 \notin \operatorname{span}(a_1, \ldots, a_d).$$

If so, j is unique.

Proof The conditions are certainly necessary, since if j is a Clifford mapping then $1 \notin \operatorname{span}(a_1, \ldots, a_d)$, $a_i^2 = j(e_i)^2 = -q(e_i)$, and $a_i a_j + a_j a_i = j(e_i)j(e_j) + j(e_j)j(e_i) = 0$, for $i \neq j$, since $e_i \perp e_j$.

Conversely, if they are satisfied, and $x = x_1 e_1 + \cdots + x_j e_j \in E$, set $j(x) = x_1 a_1 + \cdots + x_j a_j$. Then j is a linear mapping of E into A, $1 \notin j(E)$ and

$$j(x)^2 = \sum_{i=1}^{d} x_1^2 a_i^2 + \sum_{1 \leq i < j \leq d} x_i x_j (a_i a_j + a_j a_i) = -\sum_{j=1}^{d} x_i^2 q(e_i) = -q(x),$$

so that j is a Clifford mapping.

Uniqueness follows from the fact that $E = \operatorname{span}(e_1, \ldots, e_d)$. $\square$

A Clifford algebra $\mathcal{A}(E, q)$ is said to be *universal* if whenever $T \in L(E, F)$ is an isometry of (E, q) into (F, r) and $B(F, r)$ is a Clifford algebra for (F, r) then T extends to an algebra homomorphism

$\tilde{T} : \mathcal{A}(E, q) \to \mathcal{B}(F, r)$:

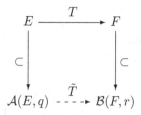

Since $\mathcal{A}(E, q)$ is generated by E, $\tilde{T}$ is then unique. Applying this to the identity mapping on E, it follows that a universal Clifford algebra is essentially unique.

Suppose that A is a Clifford algebra for (E, q). We set $\Omega = \Omega_d = \{1, \dots, d\}$. If $C = \{i_1, \dots, i_k\}$, with $1 \leq i_1 < \cdots < i_k \leq d$, then we define the element e_C of A to be the product $e_{i_1} \dots e_{i_k}$ (we set $e_\emptyset = 1$). Note that if $|C| > 1$ then e_C depends on the ordering of the set $\{1, \dots, d\}$. The element $e_\Omega = e_1 \dots e_d$ will be particularly important.

Proposition 5.1.1 *Suppose that A is a Clifford algebra for (E, q). Then $A = \mathrm{span}(P)$, where $P = \{e_C : C \subseteq \Omega\}$. If P is linearly independent (and therefore a basis for A) then A is universal.*

Proof Since $e_i^2 = 0, 1$ or -1 and $e_i e_j = -e_j e_i$ for $i \neq j$, it follows that if $e_C, e_D \in P$ then either $e_C e_D = 0$ or $e_C e_D = \pm e_{C \Delta D}$, where $C \Delta D$ is the symmetric difference $(C \setminus D) \cup (D \setminus C)$. Thus $P \cup (-P)$ is closed under multiplication, and so spans A. If P is linearly independent and $T \in L(E, F)$ is an isometry of (E, q) into (F, r) and $\mathcal{B}(F, r)$ is a Clifford algebra for (F, r), then we can extend T to an algebra homomorphism $\tilde{T} : A \to B$ by setting

$$\tilde{T}(e_{i_1} \dots e_{i_k}) = T(e_{i_1}) \dots T(e_{i_k})$$

and extending by linearity. $\square$

Corollary 5.1.1 *If $\dim A = 2^d$ then A is universal.*

5.2 Existence

Theorem 5.2.1 *If (E, q) is a quadratic space, there exists a universal Clifford algebra $\mathcal{A}(E, q)$.*

We shall give more than one proof of this important theorem. The proof given here uses creation and annihilation operators.

Proof We show that we can take $\mathcal{A}(E,q)$ to be a subalgebra of the algebra $L(\bigwedge^* E)$ of linear operators on $\bigwedge^* E$. If $x \in E$, let $j(x) = m_x - \delta_x$, where m_x is the creation operator and δ_x is the annihilation operator corresponding to x. Then

$$j(x)^2 = m_x^2 - m_x \delta_x - \delta_x m_x + \delta_x^2 = -(m_x \delta_x + \delta_x m_x) = -q(x)1,$$

by Theorem 4.1.1. Since $j(x)1 = x$, $1 \notin j(E)$. Thus j is a Clifford mapping of (E,q) into $L(\bigwedge^* E)$. We take $\mathcal{A}(E,q)$ to be the unital algebra generated by $j(E)$, and identify E with $j(E)$. It remains to show that $\mathcal{A}(E,q)$ is universal. We show that the set $\{e_C : C \subseteq \Omega\}$ is linearly independent. First, $e_\emptyset(1) = 1$. Next, we show by induction on $k = |C|$ that if $C = \{j_1 < \cdots < j_k\}$ then $e_C(1) = e_{j_1} \wedge \cdots \wedge e_{j_k}$. The result is true if $k = 1$. Suppose that it it is true for k and that $C = \{j_0 < \cdots < j_k\}$. Then

$$m_{e_{j_0}}(e_{j_1} \ldots e_{j_k})(1) = m_{e_{j_0}}(e_{j_1} \wedge \cdots \wedge e_{j_k}) = e_{j_0} \wedge e_{j_1} \wedge \cdots \wedge e_{j_k},$$
$$\delta_{e_{j_0}}(e_{j_1} \ldots e_{j_k})(1) = \delta_{e_{j_0}}(e_{j_1} \wedge \cdots \wedge e_{j_k}) = 0.$$

Thus the elements $\{e_C(1) : C \subseteq \Omega_d\}$ are linearly independent in $\bigwedge^* E$; this implies that the operators $\{e_C : C \subseteq \Omega_d\}$ are linearly independent in $L(\bigwedge^* E)$. $\square$

Corollary 5.2.1 *If A is a universal Clifford algebra for (E,q) then $\dim A = 2^d$, where $d = \dim E$. Thus a Clifford algebra A for (E,q) is universal if and only if $\dim A = 2^d$, and if and only if the set $P = \{e_C : C \subseteq \Omega\}$ is a linearly independent subset of A.*

We denote a universal Clifford algebra for (E,q) by $\mathcal{A}(E,q)$. We write $\mathcal{A}_d$ for $\mathcal{A}(\mathbf{R}_d)$ and $\mathcal{A}_{p,m}$ for $\mathcal{A}(\mathbf{R}_{p,m})$.

We can consider $\mathcal{A}(E,q)$ as a quotient of $\otimes^* E$. Let i be the inclusion mapping $E \to \otimes^* E$, and let us denote the Clifford mapping from E to a universal Clifford algebra $\mathcal{A}(E,q)$ by j_A. By Theorem 3.2.1, there is a unique algebra homomorphism k_A from $\otimes^* E$ into $\mathcal{A}(E,q)$ such that $k_A \circ i = j_A$; since E generates $\mathcal{A}(E,q)$, k_A is surjective. Let I_q be the ideal in $\otimes^* E$ generated by the elements $x \otimes x + q(x)1$, let $\mathcal{C}(E,q) = (\otimes^* E)/I_q$ be the quotient algebra, and let $\pi : \otimes^* E \to \mathcal{C}(E,q)$ be the quotient mapping. Let us set $j_B = \pi \circ i$. Then I_q is in the null-space of k_A, and so there exists a unique algebra homomorphism $J : \mathcal{C}(E,q) \to \mathcal{A}(E,q)$ such that $k_A = J \circ \pi$. Then $j_A = k_A \circ i = J \circ \pi \circ i = J \circ j_B$. Again, J is surjective. If $x \in E$, then $J(j_B(x)) = j_A(x)$, so that $1 \notin j_B(E)$. Further,

$$(j_B(x))^2 = \pi(x \otimes x) = \pi(x \otimes x + q(x)1) - \pi(q(x)1) = -q(x),$$

so that j_B is a Clifford mapping of E into $\mathcal{C}(E,q)$. Since $i(E)$ generates $\otimes^* E$, $j_B(E) = \pi(i(E))$ generates $\mathcal{C}(E,q)$, and so $\mathcal{C}(E,q)$ is a Clifford algebra for (E,q). Since $\mathcal{A}(E,q)$ is universal, there exists a unique unital algebra homomorphism $\rho : \mathcal{A}(E,q) \to \mathcal{C}(E,q)$ such that $\rho \circ j_A = j_B$. It follows that J is an algebra isomorphism of $\mathcal{C}(E,q)$ onto $\mathcal{A}(E,q)$, with inverse ρ. Thus $\mathcal{C}(E,q)$ is a universal Clifford algebra for (E,q).

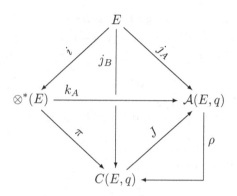

5.3 Three involutions

Conjugation plays an important part in complex analysis. The same is true in Clifford algebras; but here the situation is rather more complicated.

Suppose that (E,q) is a quadratic space, with universal Clifford algebra $\mathcal{A} = \mathcal{A}(E,q)$. Let $m(x) = -x$, for $x \in E$. Then m is an isometry of E, and so by universality we have:

$$
\begin{array}{ccc}
E & \xrightarrow{\;\;m\;\;} & E \\[2pt]
\scriptstyle\subset\big\downarrow & & \big\downarrow\scriptstyle\subset \\[2pt]
\mathcal{A}(E,q) & \dashrightarrow{\;\tilde{m}\;} & \mathcal{A}(E,q)
\end{array}
$$

We write a' for $\tilde{m}(a)$. Clearly $a \to a'$ is an automorphism, and $a'' = a$ (we have an *involution*). Also, $e'_C = (-1)^{|C|} e_C$. This involution is called the *principal* involution. We set

$$\mathcal{A}^+ = \{a : a = a'\}, \qquad \mathcal{A}^- = \{a : a = -a'\}.$$

Then $\mathcal{A} = \mathcal{A}^+ \oplus \mathcal{A}^-$, and $\mathcal{A}^+$ is a subalgebra of $\mathcal{A}$, the *even Clifford algebra.* Further,

$$\mathcal{A}^+.\mathcal{A}^+ = \mathcal{A}^-.\mathcal{A}^- = \mathcal{A}^+, \quad \mathcal{A}^+.\mathcal{A}^- = \mathcal{A}^-.\mathcal{A}^+ = \mathcal{A}^-,$$

so that $\mathcal{A}$ is a super-algebra. This super-algebra property is fundamentally important. Note that $e_C \in \mathcal{A}^+$ if $|C|$ is even, and $e_C \in \mathcal{A}^-$ if $|C|$ is odd. In particular, $j(E) \subseteq \mathcal{A}^-$.

Theorem 5.3.1 *Suppose that the quadratic space (E,q) is the orthogonal direct sum $(E_1,q_1) \oplus (E_2,q_2)$ of quadratic spaces. Then $\mathcal{A}(E,q) \cong \mathcal{A}(E_1,q_1) \otimes_g \mathcal{A}(E_2,q_2)$.*

Proof Let $\mathcal{G}$ be the graded product $\mathcal{A}(E_1,q_1) \otimes_g \mathcal{A}(E_2,q_2)$. If $x_1 \in E_1$ and $x_2 \in E_2$, define $j(x_1+x_2) = (x_1 \otimes_g 1)+(1 \otimes_g x_2)$. Then j is a one-one linear mapping of E into $\mathcal{G}$, and $1 \notin j(E)$. Since $x_1 \in \mathcal{A}^-(E_1,q_1)$ and $x_2 \in \mathcal{A}^-(E_2,q_2)$,

$$(j(x_1+x_2))^2 = -q_1(x_1) + x_1 \otimes_g x_2 - x_1 \otimes_g x_2 - q_2(x_2) = -q(x).$$

Thus j is a Clifford mapping of (E,q) into $\mathcal{G}$. Further, $j(E_1+E_2)$ generates $\mathcal{G}$, and

$$\dim \mathcal{A} = \dim \mathcal{A}(E_1,q_1).\dim \mathcal{A}(E_2,q_2) = 2^{d_1}2^{d_2} = 2^d,$$

so that $\mathcal{G}$ is a universal Clifford algebra for (E,q).   □

Corollary 5.3.1 *If (E,q) is a d-dimensional quadratic space, $\mathcal{A}(E,q)$ is isomorphic to the graded tensor product of d two-dimensional graded algebras.*

This theorem and its corollary are theoretically interesting, but are not very useful in practice, since the construction of graded tensor products is not a straightforward matter. In practice, it is more useful to construct ordinary tensor products, as we shall see in subsequent chapters.

We can however use this theorem to give a second proof of Theorem 5.2.1. This uses induction on $d = \dim E$. The result is true when $d = 1$, as the examples in the previous section show. Assume that the result is true for spaces of dimension less than $\dim E = d$, where $d > 1$. We can write E as an orthogonal direct sum $E_1 \oplus E_2$, with $\dim E_1 = d_1 < d$ and $\dim E_2 = d_2 < d$. Let q_1 and q_2 be the restrictions of q to E_1 and E_2. By the inductive hypothesis, there exist universal Clifford algebras $\mathcal{A}(E_1,q_1)$ and $\mathcal{A}(E_2,q_2)$. Let $\mathcal{G}$ be the graded product $\mathcal{A}(E_1,q_1) \otimes_g \mathcal{A}(E_2,q_2)$. If $x_1 \in E_1$ and $x_2 \in E_2$, define $j(x_1+x_2) = (x_1 \otimes_g 1)+(1 \otimes_g x_2)$. Then j is a one-one linear mapping of E into $\mathcal{G}$, and $1 \notin j(E)$. Since $x_1 \in \mathcal{A}^-(E_1, q_1)$ and $x_2 \in \mathcal{A}^-(E_2, q_2)$,

$$(j(x_1 + x_2))^2 = -q_1(x_1) + x_1 \otimes_g x_2 - x_1 \otimes_g x_2 - q_2(x_2) = -q(x).$$

Further, $j(E_1 + E_2)$ generates $\mathcal{G}$, and

$$\dim \mathcal{G} = \dim \mathcal{A}(E_1, q_1). \dim \mathcal{A}(E_2, q_2) = 2^{d_1} 2^{d_2} = 2^d,$$

so that $\mathcal{G}$ is a universal Clifford algebra for (E, q).

Note that the theorem shows that

$$\mathcal{A}_2 \cong \mathcal{A}_1 \otimes_g \mathcal{A}_1 \cong \mathbf{C} \times_g \mathbf{C} \cong \mathbf{H}.$$

We can also see this directly. Let $j((x_1, x_2)) = x_1 \boldsymbol{i} + x_2 \boldsymbol{j}$. Then j is an injective linear mapping of $\mathbf{R}_2$ into $\mathbf{H}$, $1 \notin j(\mathbf{R}_2)$ and

$$(j((x_1, x_2)))^2 = x_1^2 \boldsymbol{i}^2 + x_1 x_2 (\boldsymbol{ij} + \boldsymbol{ji}) + x_2^2 \boldsymbol{j}^2 = -x_1^2 - x_2^2.$$

Thus j is a Clifford mapping of $\mathbf{R}_2$ into $\mathbf{H}$, and so $\mathbf{H}$ is a Clifford algebra for the Euclidean space $\mathbf{R}_2$. Since $\dim \mathbf{H} = 4 = 2^2$, $\mathbf{H}$ is a universal Clifford algebra. Note that $\mathcal{A}_2^+ = \mathrm{span}(1, \boldsymbol{k})$; this appearance of the quaternions is rather unsatisfactory, since $\boldsymbol{i}, \boldsymbol{j}$ and $\boldsymbol{k}$ appear in an unsymmetric way.

Recall that if A is an algebra, then the *opposite algebra* A^{opp} is the vector space A with a new multiplication defined by $x \circ y = yx$. It is an algebra. If $\mathcal{A}(E, q)$ is a universal Clifford algebra for (E, q), with Clifford mapping j, then $j : (E, q) \to \mathcal{A}(E, q)^{opp}$ is also a Clifford mapping, and so $\mathcal{A}(E, q)^{opp}$ is also a Clifford algebra for (E, q). It is universal, since it has the same dimension as $\mathcal{A}(E, q)$. We have:

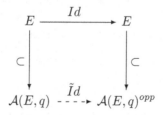

where Id is the identity mapping. We write a^* for $\tilde{Id}(a)$. Then $a^* b^* = b^* \circ a^* = (ba)^*$. Again, $a^{**} = a$, and so we have an involutory *anti-automorphism* of $\mathcal{A}(E, q)$, called the *principal anti-involution*, or *reversal*. If $C = \{e_{j_1}, \ldots, e_{j_k}\}$ with $1 \le j_1 < \cdots < j_k \le d$, then

$$e_C^* = e_{j_k} \ldots e_{j_1} = (-1)^{|C|(|C|-1)/2} e_C,$$

since e_C^* can be obtained from e_C by making $|C|(|C| - 1)/2$ transpositions.

Finally, we can consider:

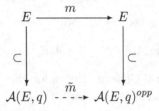

where $m(x) = -x$. We write $\tilde{m}(a) = \bar{a}$. Clearly $\bar{a} = (a')^* = (a^*)'$, $\bar{e}_C = (-1)^{|C|(|C|+1)/2} e_A$, and $\bar{\bar{a}} = a$. Further, $\overline{ab} = \bar{b}.\bar{a}$, so that the mapping $a \to \bar{a}$ is an anti-automorphism of $\mathcal{A}$, called *conjugation*.

If $e_1, \ldots, e_d$ is a standard orthogonal basis for E, then $e_C' = (-1)^{\alpha(C)} e_C$, $e_C^* = (-1)^{\beta(C)} e_C$ and $\overline{e_C} = (-1)^{\gamma(C)} e_C$, where α, β and γ are given by the following table:

$|C| \pmod 4$	0	1	2	3
$\alpha(C)$	0	1	0	1
$\beta(C)$	0	0	1	1
$\gamma(C)$	0	1	1	0

If F is a linear subspace of (E, q), then the inclusion mapping $F \to \mathcal{A}(E, q)$ is a Clifford mapping, and therefore extends to an isomorphism of $\mathcal{A}(F, q)$ into $\mathcal{A}(E, q)$. This mapping is called the *canonical inclusion* and will be denoted by $\tilde{\imath}$. Note that $(\tilde{\imath}(a))' = \tilde{\imath}(a')$, $\overline{\tilde{\imath}(a)} = \tilde{\imath}\bar{a}$ and $(\tilde{\imath}(a))^* = \tilde{\imath}(a^*)$.

Exercises

1. Suppose that F is a linear subspace of (E, q). Verify that the canonical inclusion $\mathcal{A}(F, q) \to \mathcal{A}(E, q)$ is injective.
2. Let G denote the group of automorphisms T of $\mathcal{A}(E, q)$ for which $T(E) \subseteq E$. Show that if $T \in G$ then the restriction $T_{|E}$ of T to E is an isometry of (E, q), and that the mapping $T \to T_{|E}$ is an isomorphism of G onto $O(E, q)$.

5.4 Centralizers, and the centre

In this section, we suppose that (E, q) is a regular quadratic space.
Recall that the *centralizer* $C_A(B)$ of a subset B of an algebra A is

$$C_A(B) = \{a \in A : ab = ba \text{ for all } b \in B\},$$

and that the *centre* $Z(A)$ of A is the centralizer $C_A(A)$:

$$Z(A)) = \{a \in A : ab = ba \text{ for all } b \in A\}.$$

Proposition 5.4.1 *Let* $\mathcal{A} = \mathcal{A}(E, q)$ *be a universal Clifford algebra for a regular quadratic space* (E, q).
  (i) $C_{\mathcal{A}}(\mathcal{A}^+) = \operatorname{span}(1, e_\Omega)$.
  (ii) $Z(\mathcal{A}) = \operatorname{span}(1, e_\Omega)$ *if d is odd, and* $Z(\mathcal{A}) = \operatorname{span}(1)$ *if d is even.*
  (iii) $Z(\mathcal{A}^+) = \operatorname{span}(1)$ *if d is odd, and* $Z(\mathcal{A}^+) = \operatorname{span}(1, e_\Omega)$ *if d is even.*

Proof (i) Since $e_i e_j e_\Omega = e_\Omega e_i e_j$ and the terms $e_i e_j$ generate $\mathcal{A}^+$, $\operatorname{span}(1, e_\Omega) \subseteq C_{\mathcal{A}}(\mathcal{A}^+)$. Suppose conversely that $z \in C_{\mathcal{A}}(\mathcal{A}^+)$; we can write $z = \sum_{C \subseteq \Omega} \lambda_C e_C$. It is sufficient to show that if D is a non-empty proper subset of Ω then $\lambda_D = 0$. There exist $i \in D$ and $j \notin D$. If $C \subseteq \Omega$ then $e_i e_j e_C e_i e_j = \alpha_C e_C$ where $\alpha_C = \pm 1$, and in particular $e_i e_j e_D e_i e_j = q(e_i) q(e_j) e_D$, so that $\alpha_D = q(e_i) q(e_j)$. Now

$$-q(e_i)q(e_j)z = e_i e_j z e_i e_j = \sum_{C \subseteq \Omega} \lambda_C e_i e_j e_C e_i e_j = \sum_{C \subseteq \Omega} \alpha_C \lambda_C e_C,$$

so that $\sum_{C \subseteq \Omega}(\alpha_C + q(e_i)q(e_j))\lambda_C e_C = 0$, and $(\alpha_C + q(e_i)q(e_j))\lambda_C = 0$ for $C \subseteq \Omega_d$. Since $\alpha_D = q(e_i)q(e_j)$, it follows that $\lambda_D = 0$.
  (ii) $Z(\mathcal{A}) \subseteq C_{\mathcal{A}}(\mathcal{A}^+)$. If d is odd, then $e_j e_\Omega = e_\Omega e_j$ for each j, so that $e_\Omega \in Z(\mathcal{A})$, and $Z(\mathcal{A}) = \operatorname{span}(1, e_\Omega)$. If d is even, then $e_j e_\Omega = -e_\Omega e_j$ for each j, so that $e_\Omega \notin Z(\mathcal{A})$, and $Z(\mathcal{A}) = \operatorname{span}(1)$.
  Finally (iii) follows from (i), since $e_\Omega \in \mathcal{A}^+$ if and only if d is even. $\square$

Corollary 5.4.1 *The element* e_Ω *is, up to choice of sign, independent of the choice of standard orthogonal basis. That is, the set* $\{e_\Omega, -e_\Omega\}$ *is independent of the choice of standard orthogonal basis.*

Proof It follows from the proposition that $\operatorname{span}(1, e_\Omega)$ does not depend on the choice of standard orthogonal basis. If $x \in \operatorname{span}(1, e_\Omega)$ then $x^2 = \pm 1$ if and only if $x = 1, -1, e_\Omega$ or $-e_\Omega$. $\square$

We call e_Ω a *volume element* of $\mathcal{A}$. It depends upon the orientation that we use. If we replace the standard orthogonal basis $(e_1, \ldots, e_d)$ by the standard orthogonal basis $(-e_1, e_2, \ldots, e_d)$ then e_Ω is replaced by $-e_\Omega$.

Note that $e_\Omega^2 = (-1)^\eta$, where $\eta = \frac{1}{2}d(d-1) + p$. If $e_\Omega^2 = 1$ then the algebra $A(e_\Omega) \cong \mathbf{R} \oplus \mathbf{R} = \mathbf{R}^2$, while if $e_\Omega^2 = -1$ then the algebra $A(e_\Omega) \cong \mathbf{C}$.

When is $e_\Omega^2 = 1$? We consider two cases. First suppose that $d = 2k$ is even and that $p = k + t$, $m = k - t$. Then $\eta = k(2k-1) + k + t = 2k^2 + t$. Thus $e_\Omega^2 = 1$ if and only if t is even; this is the case if and only if $p - m = 2t = 0 \pmod 4$.

Secondly, if $d = 2k + 1$ is odd and $p = k + t$, $m = k + 1 + t$ then $\eta = k(2k + 1) + k + t = 2k(k + 1) + t$. Again, $e_\Omega^2 = 1$ if and only if t is even; this is the case if and only if $p - m = 2t - 1 = 3 \pmod 4$.

We therefore we have the following table:

$p - m(\mathrm{mod}4)$	0	1	2	3
$C_\mathcal{A}(\mathcal{A}^+)$	$\mathbf{R}^2$	$\mathbf{C}$	$\mathbf{C}$	$\mathbf{R}^2$
$Z(\mathcal{A}(E,q))$	$\mathbf{R}$	$\mathbf{C}$	$\mathbf{R}$	$\mathbf{R}^2$
$Z(\mathcal{A}^+(E,q))$	$\mathbf{R}^2$	$\mathbf{R}$	$\mathbf{C}$	$\mathbf{R}$

This means that $\mathcal{A}(E,q)$ can be considered as a complex algebra in a natural way if and only if d is odd and $p - m = 1 \pmod 4$. Similarly, $\mathcal{A}^+(E,q)$ can be considered as a complex algebra in a natural way if and only if d is even and $p - m = 2 \pmod 4$.

5.5 Simplicity

Are there any non-universal Clifford algebras? If A is a Clifford algebra for (E, q) then we have a diagram

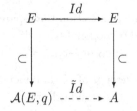

and $\tilde{I}d$ is an algebra homomorphism of $\mathcal{A}(E,q)$ onto A. If $\mathcal{A}(E,q)$ is simple, then $\tilde{I}d$ is an algebra isomorphism of $\mathcal{A}(E,q)$ onto A, and so (E,q) has no non-universal Clifford algebra.

If (E,q) is not regular, then $e_\Omega^2 = 0$, and $\mathcal{A}(E,q)e_\Omega$ is a proper ideal in $\mathcal{A}(E,q)$. Let $\mathcal{B}(E,q)$ be the quotient algebra $\mathcal{A}(E,q)/\mathcal{A}(E,q)e_\Omega$, and let $q : \mathcal{A}(E,q) \to \mathcal{B}(E,q)$ be the quotient mapping. If $\dim E > 1$ and $j : (E,q) \to \mathcal{A}(E,q)$ is the Clifford mapping, then $q \circ j$ is a Clifford mapping from (E,q) into $\mathcal{B}(E,q)$, so that $\mathcal{B}(E,q)$ is a non-universal Clifford algebra for (E,q).

Suppose that (E,q) is regular, with signature (p,m). Then $(A,q) \cong \mathcal{A}_{p,m}$, and so it is sufficient to consider the algebras $\mathcal{A}_{p,m}$.

Theorem 5.5.1 *If $p - m \neq 3$ (mod 4) then $\mathcal{A}_{p,m}$ is simple, so that $\mathbf{R}_{p,m}$ has no non-universal Clifford algebras.*

Proof We can suppose that $d \geq 2$. Suppose that I is a non-zero ideal in $\mathcal{A}_{p,m}$. Let x be a non-zero element in I with a minimal number of non-zero coefficients in its expansion with respect to the basis $\{e_C : C \subseteq \Omega\}$. By multiplying by an e_C and scaling, we can suppose that

$$x = 1 + \sum_{C \in R} \lambda_C e_C,$$

where R is a set of nonempty subsets of Ω. We shall show that R is the empty set. Suppose that $B \in R$ and that $B \neq \Omega$, so that there exist $i \in B$ and $j \notin B$. Then

$$e_i e_j x e_i e_j = -q(e_i)q(e_j) + \sum_{C \in R} \mu_C e_C,$$

where $\mu_C = \pm q(e_i)q(e_j)\lambda_C$, and where $\mu_B = q(e_i)q(e_j)\lambda_B$. Consequently $q(e_i)q(e_j)x - e_i e_j x e_i e_j$ is a non-zero element of I with fewer non-zero terms, giving a contradiction. Thus $x = 1 + \lambda_\Omega e_\Omega$.

Suppose that d is even. Then $e_1 x e_1 = -q(e_1)(1 - \lambda_\Omega e_\Omega)$, so that $q(e_1)x - e_1 x e_1 = 2q(e_1)1 \in I$ and $I = \mathcal{A}_{p,m}$. If $p - m = 1$ (mod 4), then $e_\Omega^2 = -1$, so that $x(1 - \lambda_\Omega e_\Omega) = (1 + \lambda_\Omega^2)1$, and again $I = \mathcal{A}_{p,m}$. $\square$

We have corresponding results for even Clifford algebras. Combining this theorem with Theorem 6.4.1, we have the following.

Corollary 5.5.1 *If $p \neq m$ (mod 4) then $\mathcal{A}_{p,m}^+$ is simple.*

What happens when $(E,q) = \mathbf{R}_{p,m}$, with $p - m = 3$ (mod 4)? In this case, $f = \frac{1}{2}(1 + e_\Omega)$ and $g = \frac{1}{2}(1 - e_\Omega)$ are idempotents in the centre of the universal Clifford algebra $\mathcal{A} = \mathcal{A}_{p,m}$ satisfying $f + g = 0$ and $f + g = 1$, so that we have a decomposition of $\mathcal{A}$ as the direct sum of two ideals $\mathcal{A}f \oplus \mathcal{A}g$. $\mathcal{A}f$ and $\mathcal{A}g$ are unital algebras, with identity elements f and g respectively. The mapping $m_f : a \to af$ is an algebra homomorphism of $\mathcal{A}$ onto $\mathcal{A}f$ with null-space $\mathcal{A}g$. If x is a non-zero element of E then $j(x)f = \frac{1}{2}(x + xe_\Omega) \notin \text{span}(f)$, and $(j(x)f)^2 = j(x)^2 f = -q(x)f$, so that the mapping $x \to j(x)f$ is a Clifford mapping of E into $\mathcal{A}f$. $j(E)f$ generates $\mathcal{A}f$, and so $\mathcal{A}f$ is a non-universal Clifford algebra for (E, q).

Now $d = p + m$ is odd, so that if $a \in \mathcal{A}^+$ then $ae_\Omega \in \mathcal{A}^-$, and so $(af)^+ = a/2$. Thus m_f is one-one on $\mathcal{A}^+$. Since $\mathcal{A}^+$ and $\mathcal{A}f$ have the same dimension, the restriction of m_f to $\mathcal{A}^+$ is an algebra isomorphism of $\mathcal{A}^+$ onto $\mathcal{A}f$. Consequently $\mathcal{A}^+ \cong \mathcal{A}f$. The same holds for the mapping $m_g : a \to ag$ from $\mathcal{A}$ onto $\mathcal{A}g$, and so $\mathcal{A}(E, q) \cong \mathcal{A}^+(E, q) \oplus \mathcal{A}^+(E, q)$.

Note that since f is neither in $\mathcal{A}^+$ nor in $\mathcal{A}^-$, the $\mathbf{Z}_2$ grading of $\mathcal{A}$ does not transfer to $\mathcal{A}f$. Since $e'_\Omega = -e_\Omega$, the principal involution maps $\mathcal{A}f$ isomorphically onto $\mathcal{A}g$, and maps $\mathcal{A}g$ isomorphically onto $\mathcal{A}f$.

Suppose that $\mathcal{B}$ is a Clifford algebra for $(E, q) \cong \mathbf{R}_{p,m}$, where $p - m = 3$ (mod 4). Then we have a diagram

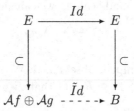

Now $\mathcal{A}f \cong \mathcal{A}^+(E, q)$, and $\mathcal{A}^+(E, q)$ is simple, by Corollary 5.5.1, so that either $\tilde{I}d(\mathcal{A}f) = 0$ or $\tilde{I}d$ is injective on $\mathcal{A}f$; a similar result holds for $\mathcal{A}g$. From this it follows that either $\mathcal{B}$ is universal, or $\mathcal{B} \cong \mathcal{A}f$. Summing up, we have the following theorem.

Theorem 5.5.2 *If $p - m = 3$ (mod 4) then $\mathcal{A}_{p,m}$ is not simple, and*

$$\mathcal{A}_{p,m} \cong \mathcal{A}_{p,m}^+ \oplus \mathcal{A}_{p,m}^+.$$

There exists a non-universal Clifford algebra $\mathcal{B}_{p,m}$ for $\mathbf{R}_{p,m}$, and any such algebra is isomorphic to $\mathcal{A}_{p,m}^+$.

When $p - m = 3$ (mod 4), it is sometimes easier to construct the non-universal Clifford algebra $\mathcal{B}_{p,m}$ than to construct the universal Clifford algebra $\mathcal{A}_{p,m}$; it is then easy to construct the Clifford algebra $\mathcal{A}_{p,m}$ from the Clifford algebra $\mathcal{B}_{p,m}$. Suppose that $k : \mathbf{R}_{p,m} \to \mathcal{B}_{p,m}$ is a Clifford mapping from $\mathbf{R}_{p,m}$ into $\mathcal{B}_{p,m}$. If $e_\Omega^\mathcal{B}$ is the corresponding volume element, then $e_\Omega^\mathcal{B} = \pm I$. The mapping $j(x) = (k(x), -k(x))$ from $\mathbf{R}_{p,m}$

into $\mathcal{B}_{p,m} \oplus \mathcal{B}_{p,m}$ is then a Clifford mapping, which extends to an algebra homomorphism $j : \mathcal{A}_{p,m} \to \mathcal{B}_{p,m} \oplus \mathcal{B}_{p,m}$. Let $e_\Omega^{\mathcal{A}}$ be the volume element in $\mathcal{A}_{p,m}$. Since $d = p + m$ is odd, $j(e_\Omega^{\mathcal{A}}) = (e_\Omega^{\mathcal{B}}, -e_\Omega^{\mathcal{B}})$, and so j is injective. Since $\dim \mathcal{A}_{p,m} = 2 \dim \mathcal{B}_{p,m}$, it follows that j is an isomorphism of $\mathcal{A}_{p,m}$ onto $\mathcal{B}_{p,m} \oplus \mathcal{B}_{p,m}$. Thus, using the Clifford mapping j, we can take $\mathcal{B}_{p,m} \oplus \mathcal{B}_{p,m}$ to be a universal Clifford algebra for $\mathbf{R}_{p,m}$. Note that then

$$\mathcal{A}_{p,m}^+ = \{(b,b) : b \in \mathcal{B}_{p,m}\} \cong \mathcal{B}_{p,m}$$
$$\mathcal{A}_{p,m}^- = \{(b,-b) : b \in \mathcal{B}_{p,m}\}.$$

Corollary 5.5.2 *If A is a Clifford algebra for a regular quadratic space (E,q) and if $a \in A$ has a left inverse b, then a is invertible, with inverse b.*

Proof If A is simple, this is an immediate consequence of Corollary 2.7.2. If A is not simple, then A is universal, and there exists an isomorphism $\pi : A \to A^+ \oplus A^+$. Let $\pi(a) = (a_1, a_2)$ and let $\pi(b) = (b_1, b_2)$. Then b_i is a left inverse of a_i in A^+, for $i = 1, 2$. Since A^+ is simple, a_i is invertible in A^+, with inverse b_i, for $i = 1, 2$. Thus a is invertible, with inverse b. $\quad\square$

Exercises

1. Show that the even Clifford algebras $\mathcal{A}^+(E,q)$ and $\mathcal{A}^+(E,-q)$ are isomorphic.

2. Suppose that $p - m = 3 \pmod 4$. If $x \in \mathbf{R}_{p,m}$, let $m_\Omega(x) = x e_\Omega$. Show that m_Ω is a Clifford map from $\mathbf{R}_{p,m}$ into $\mathcal{A}_{p,m}$, and let m_Ω also denote the corresponding algebra homomorphism from $\mathcal{A}_{p,m}$ into $\mathcal{A}_{p,m}$. What are the fixed points of m_Ω? What is the image of m_Ω? What is the null-space of m_Ω?

3. Suppose that (E, q) has signature (p, m), where $p - m = 3 \pmod 4$. Let $i : (E, q) \to (E, -q)$ be the identity mapping, and let $n_\Omega : (E, q) \to \mathcal{A}(E, -q)$ be defined by $n_\Omega(x) = i(x)e_\Omega$, where e_Ω is a volume element in $\mathcal{A}(E, -q)$. Show that n_Ω is a Clifford map from (E, q) into $\mathcal{A}(E, -q)$. Let n_Ω also denote the corresponding algebra homomorphism from $\mathcal{A}(E, q)$ into $\mathcal{A}(E, -q)$. What is the image of n_Ω? What is the null-space of n_Ω?

5.6 The trace and quadratic form on $\mathcal{A}(E, q)$

Suppose that $\mathcal{A} = \mathcal{A}(E, q)$ is a universal Clifford algebra for a regular quadratic space (E, q), so that the set $\{e_C : C \subseteq \Omega_d\}$ is a basis for $\mathcal{A}$. Although the definition of the normalized trace τ_n on $\mathcal{A}$ does not depend on the choice of basis, we can use this basis to obtain properties of τ_n. If $C \neq \emptyset$ and $D \subseteq \Omega_d$ then

$$l_{e_C}(e_D) = e_C e_D = \pm e_{C \Delta D} \neq \pm e_D,$$

(where $C \Delta D = (C \setminus D) \cup (D \setminus C)$), so that the diagonal terms of the matrix representing l_{e_C} all vanish, and $\tau_n(e_C) = 0$. Thus if $a = \sum_{C \subseteq \Omega_d} \alpha_C e_C$ then $\tau_n(a) = \alpha_\emptyset$. Note also that

$$\tau_n(a) = \tau_n(a') = \tau_n(a^*) = \tau_n(\bar{a}).$$

In particular, if $a \in \mathcal{A}^-(E, q)$ then $\tau_n(a) = 0$.

Theorem 5.6.1 *The normalized trace τ_n is the only linear functional ϕ on $\mathcal{A} = \mathcal{A}(E, q)$ which satisfies the conditions*
  (i) $\phi(1) = 1$;
  (ii) $\phi(ab) = \phi(ba)$ for all $a, b \in \mathcal{A}$;
  (iii) $\phi(a) = \phi(a')$ for all $a \in \mathcal{A}$.

Proof We have seen that τ_n satisfies the conditions. Suppose that ϕ satisfies the conditions. If $|C|$ is odd then $e'_C = -e_C$, so that $\phi(e_C) = -\phi(e_C)$, and $\phi(e_C) = 0$. Suppose that $|C|$ is even and that $C \neq \emptyset$. Let j be the least element of C and let $D = C \setminus \{j\}$, so that $e_C = e_j e_D = -e_D e_j$. Then

$$\phi(e_C) = \phi(e_j e_D) = -\phi(e_D e_j) = -\phi(e_j e_D) = -\phi(e_C),$$

so that $\phi(e_C) = 0$. Thus $\phi = \tau_n$. □

We now use the normalized trace on $\mathcal{A}(E,q)$ to define a quadratic form on $\mathcal{A}(E,q)$. If $a,b \in \mathcal{A}$, let $B(a,b) = \tau_n(a\bar{b})$. Then B is a bilinear form on $\mathcal{A}(E,q)$. Since

$$B(a,b) = \tau_n(a\bar{b}) = \tau_n(\overline{\overline{a}b}) = \tau_n(b\bar{a}) = B(b,a),$$

B is symmetric, and so it defines a quadratic form Q on $\mathcal{A}(E,q)$: $Q(a) = \tau_n(a\bar{a}) = \tau_n(\bar{a}a)$. Since $e_C \bar{e}_C = \prod_{j \in C} q(e_j)$, it follows that $Q(e_C) = \prod_{j \in C} q(e_j)$. In particular, if $x \in E$ then $Q(x) = q(x)$. If (E,q) is regular, then Q is non-singular, and the set $\{e_C : C \subseteq \Omega_d\}$ is a standard orthogonal basis for $\mathcal{A}(E,q)$. If (E,q) is positive definite, then so is $(\mathcal{A},Q)$. On the other hand, if (E,q) is regular, but not positive definite, and if $A \subset \{1,\dots,d-1\}$ then $Q(e_{A\cup\{d\}}) = Q(A)Q(e_d) = -Q(e_A)$; it follows easily from this that $(\mathcal{A}(E,q),Q)$ is hyperbolic.

The quadratic form Q enables us to decompose $\mathcal{A}(E,q)$ as a direct sum of certain linear subspaces. Let $(e_1,\dots,e_d)$ be a standard orthogonal basis for E, and let

$$A_j = \mathrm{span}\{e_C : |C| = j\}, \quad B_j = \mathrm{span}\{e_C : |C| \le j\}, \text{ for } 0 \le j \le d,$$

with the convention that $A_0 = B_0 = \mathbf{R}$. Then

$$B_j = \mathrm{span}\{a \in \mathcal{A}(E,q) : a \text{ is the product of at most } j \text{ vectors}\},$$

so that B_j does not depend upon the choice of standard orthogonal basis. But $A_j = B_j \cap B_{j-1}^{\perp}$, and so A_j also does not depend upon the choice of standard orthogonal basis. We have an orthogonal decomposition $\mathcal{A}(E,q) = A_0 \oplus A_1 \oplus \cdots \oplus A_d$: $A_1 = E$ and $A_d = \mathrm{span}(e_\Omega)$. Elements of A_2 are called *bivectors*, elements of A_2 are called *trivectors*, and so on. (A word of caution here: some authors use the term bivector to mean the product of two orthogonal vectors.)

5.7 The group $\mathcal{G}(E,q)$ of invertible elements of $\mathcal{A}(E,q)$

Suppose that (E,q) is a regular quadratic space. We denote the group of invertible elements of the universal Clifford algebra $\mathcal{A}(E,q)$ by $\mathcal{G}(E,q)$. The anisotropic elements of E are contained in $\mathcal{G}(E,q)$.

If $a \in \mathcal{A}(E,q)$, let $\Delta(a) = a\bar{a}$; $\Delta(a)$ is the *quadratic norm* on $\mathcal{A}(E,q)$. Note that $Q(a) = \tau_n(\Delta(a))$. The element $\Delta(a)$ provides a useful test to determine whether a is invertible.

Proposition 5.7.1 *If $a \in \mathcal{A}(E, q)$ then $a \in \mathcal{G}(E, q)$ if and only if $\Delta(a) \in \mathcal{G}(E, q)$.*

Proof If $a \in \mathcal{G}(E, q)$ then $\bar{a} \in \mathcal{G}(E, q)$, so that $\Delta(a) \in \mathcal{G}(E, q)$. Conversely if $\Delta(a) \in \mathcal{G}(E, q)$ then $(\Delta(a))^{-1}\bar{a}a = I$ so that a has a left inverse. This implies that $a \in \mathcal{G}(E, q)$, by Corollary 5.5.2. $\square$

For example, if $x = a1 + b\boldsymbol{i} + c\boldsymbol{j} + d\boldsymbol{k} \in \mathbf{H}$ then, as in Section 2.3, $\Delta(x) = a^2 + b^2 + c^2 + d^2$, and so $x \in \mathcal{G}(\mathbf{R}_2)$ if and only if $x \neq 0$.

If $\lambda 1 + x \in \mathbf{R} \oplus E$ then $\Delta(\lambda 1 + x) = (\lambda 1 - x)(\lambda 1 + x) = \lambda^2 + q(x)$, so that $\lambda 1 + x \in \mathcal{G}(E, q)$ if and only if $q(x) \neq -\lambda^2$; in particular, if (E, q) is Euclidean then the non-zero elements of $\mathbf{R} \oplus E$ are invertible. Suppose that $\lambda 1 + x$ is an invertible element of $\mathbf{R} \oplus E$. Then if $a \in \mathcal{A}(E, q)$,

$$Q((\lambda 1 + x)a) = \tau_n(\bar{a}(\lambda 1 - x)(\lambda 1 + x)a)$$
$$= \tau_n((\lambda^2 + q(x))\bar{a}a) = (\lambda^2 + q(x))Q(a),$$

so that in the left regular representation of $\mathcal{A}(E, q)$, $l_{\lambda 1 + x}$ is a multiple of an isometry of $\mathcal{A}(E, q)$.

We can generalize this. Let

$$\mathcal{N} = \mathcal{N}(E, q) = \{a : \Delta(a) \in \mathbf{R}\},$$
$$\mathcal{N}_* = \{a : \Delta(a) \in \mathbf{R}^*\} = \{a \in \mathcal{N} : \Delta(a) \neq 0\},$$
$$\mathcal{N}_{\pm 1} = \{a \in \mathcal{N} : \Delta(a) = \pm 1\},$$
$$\mathcal{N}_1 = \{a \in \mathcal{N} : \Delta(a) = 1\}.$$

We set $\mathcal{N}^+ = \mathcal{N} \cap \mathcal{A}^+(E, q)$, and use the same convention for the other terms.

If $a \in \mathcal{N}$ then $Q(a) = \tau_n(\Delta(a)) = \Delta(a)$. Note that if (E, q) is a Euclidean space then $\Delta(a) \geq 0$ for $a \in \mathcal{N}$, and $\Delta(a) > 0$ for $a \in \mathcal{N}_*$.

Proposition 5.7.2 $\mathcal{N}_*$ *is a subgroup of $\mathcal{G}(E, q)$, and Δ is a character on $\mathcal{N}_*$: $\Delta(ab) = \Delta(a)\Delta(b)$. If $a \in \mathcal{N}_*$ then a', a^* and $\bar{a}$ are in $\mathcal{N}_*$, and*

$$\Delta(a) = \Delta(a') = \Delta(a^*) = \Delta(\bar{a}).$$

Proof If $a \in \mathcal{N}_*$ then $\Delta(a) \in \mathcal{G}(E, q)$, and so $a \in \mathcal{G}(E, q)$.

If $a \in \mathcal{N}_*$ and $b \in \mathcal{A}$ then

$$\Delta(ab) = \bar{b}\bar{a}ab = \bar{b}\Delta(a)b = \Delta(a)\bar{b}b = \Delta(a)\Delta(b);$$

in particular, Δ is multiplicative on $\mathcal{N}_*$.

Further, $a(\bar{a}/\Delta(a)) = 1$, so that $a\bar{a} = \Delta(a)$, $\bar{a} \in \mathcal{N}_*$, $\Delta(a) = \Delta(\bar{a})$ and

$a^{-1} = \bar{a}/\Delta(a) \in \mathcal{N}_*$: $\mathcal{N}_*$ is a subgroup of $\mathcal{G}(E,q)$. Since

$$\Delta(a') = (\bar{a'})a' = (\bar{a}a)' = (\Delta(a))' = \Delta(a),$$

$a' \in \mathcal{N}_*$. Finally $a^* \in \mathcal{N}_*$ and $\Delta(a^*) = \Delta(a)$.   □

Proposition 5.7.3 *If $a \in \mathcal{N}$ and $b \in \mathcal{A}$ then $Q(ab) = \Delta(a)Q(b)$, so that in the left regular representation of $\mathcal{A}$, $l(a)$ is a scalar multiple of an isometry of A. Thus the mapping $a \to l(a)$ is an injective homomorphism of $\mathcal{N}_1(E,q)$ into the orthogonal group $O(\mathcal{A}(E,q),Q)$.*

Proof $Q(ab) = \tau_n(a\bar{b}\bar{b}\bar{a}) = \tau_n(\bar{a}a\bar{b}b) = \Delta(a)\tau_n(\bar{b}b) = \Delta(a)Q(b)$.   □

6

Classifying Clifford algebras

From now on, we shall only consider Clifford algebras for regular quadratic spaces.

6.1 Frobenius' theorem

We have seen that $\mathbf{H}$ is a division algebra.

Theorem 6.1.1 (Frobenius' theorem) $\mathbf{R}$, $\mathbf{C}$ *and* $\mathbf{H}$ *are the only finite-dimensional real division algebras.*

Proof Suppose that A is a finite-dimensional real division algebra, of dimension $d > 1$. We identify $\mathbf{R}$ with $\mathbf{R}.1$. We consider the set $E = \{a \in A : a^2 \in \mathbf{R}\}$ and its subsets

$$E^+ = \{a \in E : a^2 \geq 0\} \quad \text{and} \quad E^- = \{a \in E : a^2 \leq 0\}.$$

First observe that $E^+ = \mathbf{R}$. For if $a^2 = c \geq 0$, then $(a - \sqrt{c})(a + \sqrt{c}) = 0$. Since A is a division algebra, either $a - \sqrt{c} = 0$ or $a + \sqrt{c} = 0$, so that $a \in \mathbf{R}$. Conversely if $a \in \mathbf{R}$ then $a^2 \geq 0$.

Next we show that $A = \operatorname{span}(E)$. Suppose that $a \in A$. Let m_a be its minimal polynomial; that is, the monic real polynomial p of minimal degree for which $p(a) = 0$. m_a must be irreducible, since if $m_a = p_1 p_2$ then $0 = m_a(a) = p_1(a)p_2(a)$, so that $p_1(a) = 0$ or $p_2(a) = 0$. Consider m_a as a complex polynomial, and let z be a complex root. If z is real, then $m_a(x) = x - z$, so that $a = z.1 \in \mathbf{R}$. If $z = g + ih$ is not real, then $\bar{z}$ is also a root of m_a, and $(x - z)(x - \bar{z}) = (x - g)^2 + h^2$ is a real polynomial which divides m_a. Thus $m_a(x) = (x - g)^2 + h^2$. Consequently $(a - g.1)^2 = -h^2.1$ and $a - g.1 \in E^-$. Thus $a = (a - g.1) + g.1 \in \operatorname{span}(E)$.

We now show that E^- is a linear subspace of A. Clearly if $a \in E^-$

and $r \in \mathbf{R}$ then $ra \in E^-$. Suppose that a and b are linearly independent elements of E^-. We show that a, b and 1 are linearly independent. If not, and $\lambda a + \mu b = \nu.1$, with λ, μ and ν not all zero, then

$$\lambda^2 a^2 = (\nu.1 - \mu b)^2 = (\nu^2.1 + \mu^2 b^2) - 2\nu\mu b \in \mathbf{R}.$$

Since $\nu^2.1 + \mu^2 b^2 - \lambda^2 a^2 \in \mathbf{R}$, $2\nu\mu b = 0$, so that either $\nu = 0$ (so that a and b are linearly dependent) or $\mu = 0$ (so that $a \in \mathbf{R}$); in either case we have a contradiction.

Considering the minimal polynomials of $a + b$ and $a - b$, we can write

$$(a + b)^2 = r(a + b) + s.1,$$
$$(a - b)^2 = t(a - b) + u.1,$$

with $r, s, t, u \in \mathbf{R}$. Adding,

$$2a^2 + 2b^2 = (r + t)a + (r - t)b + (s + u).1.$$

Since a, b and 1 are linearly independent, $r + t = 0$ and $r - t = 0$, so that $r = 0$, and $a + b \in E$. But if $a + b \in \mathbf{R}$, then either $a + b = 0$, or a, b and $a + b$ are linearly independent. Neither is possible, and so we have a contradiction. Thus $a + b \in E^-$, E^- is a linear subspace of A, and $A = \mathbf{R}.1 \oplus E^-$.

Let $\beta(a, b) = -\frac{1}{2}(ab + ba)$, for $a, b \in E^-$. Then β is a symmetric bilinear mapping of $E^- \times E^-$ into A. But $\beta(a, b) = \frac{1}{2}(a^2 + b^2 - (a+b)^2)$, so that β takes values in $\mathbf{R}$. Let $q(a) = \beta(a, a) = -a^2$. Then q is a positive definite quadratic form on E^-, and the inclusion mapping $j : E^- \to A$ is a Clifford mapping. Since $A = \text{span}(1, E^-)$, A is a Clifford algebra for E^-. If A is universal, then $\dim A = 2^d = d + 1$, so that $d = 1$ and $A \cong \mathbf{C}$. If A is not universal, then $\dim A = 2^{d-1} = d + 1$, so that $d = 3$ and $A \cong \mathbf{H}$.   □

Corollary 6.1.1 *(i) Suppose that $d = p + m = 2k + 1$ is odd.*
If $p - m = 3 \pmod 4$ then $\mathcal{A}_{p,m} \cong M_{2^k}(\mathbf{C})$.
If $p - m = 1 \pmod 4$ then either $\mathcal{A}_{p,m} \cong M_{2^k}(\mathbf{R}) \oplus M_{2^k}(\mathbf{R})$ or $\mathcal{A}_{p,m} \cong M_{2^{k-1}}(\mathbf{H}) \oplus M_{2^{k-1}}(\mathbf{H})$.
(ii) Suppose that $d = p + m = 2k$ is even. Then $\mathcal{A}_{p,m} \cong M_{2^k}(\mathbf{R})$ or $M_{2^{k-1}}(\mathbf{H})$.

Proof Since the centre of $M_n(\mathbf{D})$ is isomorphic to the centre of $\mathbf{D}$, and since $\mathcal{A}_{p,m}$ has real dimension 2^{p+m}, these results follow from Wedderburn's theorem and from Theorems 5.5.1 and 5.5.2.   □

Although it is of interest to see how Wedderburn's theorem and Frobenius' theorem relate to the structure of Clifford algebras, we shall in fact classify the Clifford algebras without using them.

Exercise

1. Suppose that A is a unital algebra in which every non-zero element has a left inverse. Show that A is a division algebra.

6.2 Clifford algebras $\mathcal{A}(E, q)$ with $\dim E = 2$

We have shown in Corollary 5.3.1 that a universal Clifford algebra is isomorphic to the graded tensor product of two-dimensional algebras. Graded tensor products are not easy to work with; in the next section we shall show that a universal Clifford algebra for a regular quadratic space of even dimension $2k$ is isomorphic to the ordinary tensor product of k four-dimensional algebras. Here we consider the case where $k = 1$.

The algebra $\mathcal{A}_2$

We have seen in Section 6.3 that the mapping $j : \mathbf{R}_2 \to \mathbf{H}$ defined by $j(x_1 e_1 + x_2 e_2) = x_1 \boldsymbol{i} + x_2 \boldsymbol{j}$ is a Clifford mapping, which extends to an algebra isomorphism of $\mathcal{A}_2$ onto the division algebra $\mathbf{H}$ of quaternions. Then $j(e_\Omega) = \boldsymbol{i}.\boldsymbol{j} = \boldsymbol{k}$, and so $j(\mathcal{A}_2^+) = \mathrm{span}(1, \boldsymbol{k}) \cong \mathbf{C}$.

The algebra $\mathcal{A}_{1,1}$

We define a Clifford mapping from $\mathbf{R}_{1,1}$ into $M_2(\mathbf{R})$ by setting

$$j(x_1 e_1 + x e_2) = x_1 J + x_2 Q = \begin{bmatrix} 0 & -x_1 + x_2 \\ x_1 + x_2 & 0 \end{bmatrix}.$$

Then j extends to an algebra homomorphism of $\mathcal{A}_{1,1}$ into $M_2(\mathbf{R})$. Since $\mathcal{A}_{1,1}$ is simple and $\dim \mathcal{A}_{1,1} = \dim M_2(\mathbf{R}) = 4$, it follows that $\mathcal{A}_{1,1} \cong M_2(\mathbf{R})$. Note that $e_\Omega = JQ = -U$. Then

$$j(x_0 1 + x_1 e_1 + x_2 e_2 + x_\Omega e_\Omega) = \begin{bmatrix} x_0 - x_\Omega & -x_1 + x_2 \\ x_1 + x_2 & x_0 + x_\Omega \end{bmatrix}.$$

Thus

$$\mathcal{A}_{1,1}^+ = \left\{ \begin{bmatrix} a & 0 \\ 0 & d \end{bmatrix} : a, d \in \mathbf{R} \right\} \cong \mathbf{R} \oplus \mathbf{R}.$$

The algebra $\mathcal{A}_{0,2}$

Let

$$j(x_1 e_1 + x e_2) = x_1 Q + x_2 U = \begin{bmatrix} x_2 & x_1 \\ x_1 & -x_2 \end{bmatrix}.$$

Then j is a Clifford mapping of $\mathbf{R}_{0,2}$ into $M_2(\mathbf{R})$, and so generates a Clifford algebra for $\mathbf{R}_{0,2}$. Since $\mathcal{A}_{0,2}$ is simple and since $\dim \mathcal{A}_{0,2} = \dim M_2(\mathbf{R}) = 4$, it follows that $\mathcal{A}_{0,2} \cong M_2(\mathbf{R})$. Note that $j(e_\Omega) = QU = J$. Then

$$j(x_0 1 + x_1 e_1 + x_2 e_2 + x_\Omega e_\Omega) = \begin{bmatrix} x_0 + x_2 & x_1 - x_\Omega \\ x_1 + x_\Omega & x_0 - x_2 \end{bmatrix}.$$

Thus

$$j(A_{1,1}^+) = \left\{ \begin{bmatrix} x & -y \\ y & x \end{bmatrix} : x, y \in \mathbf{R} \right\} \cong \mathbf{C}.$$

Exercises

1. Let $j : \mathcal{A}_2 \to \mathbf{H}$ be the isomorphism defined above. Show that if $x = a_0 1 + x_1 e_1 + x_2 e_2 + x_\Omega e_\Omega \in \mathcal{A}_2$ then

$$j(x') = x_0 1 - x_1 i - x_2 j + x_\Omega k,$$
$$j(x^*) = x_0 1 + x_1 i + x_2 j - x_\Omega k,$$
$$j(\bar{x}) = x_0 1 - x_1 i - x_2 j - x_\Omega k,$$

and that $x\bar{x} = \bar{x}x = (x_0^2 + x_1^2 + x_2^2 + x_\Omega^2)1$.

2. Let $j : \mathcal{A}_{1,1} \to M_2(\mathbf{R})$ be the isomorphism defined above. Show that if $x \in \mathcal{A}_{1,1}$ and

$$j(x) = \begin{bmatrix} a & b \\ c & d \end{bmatrix} \in M_2(\mathbf{R})$$

then

$$j(x') = \begin{bmatrix} a & -b \\ -c & d \end{bmatrix}, \quad j(x^*) = \begin{bmatrix} d & b \\ c & a \end{bmatrix}, \quad j(\bar{x}) = \begin{bmatrix} d & -b \\ -c & a \end{bmatrix},$$

and

$$x\bar{x} = \bar{x}x = \det \begin{bmatrix} a & b \\ c & d \end{bmatrix}.I.$$

3. Let $j : \mathcal{A}_{0,2} \to M_2(\mathbf{R})$ be the isomorphism defined above. Show that if $x \in \mathcal{A}_{0,2}$ and

$$j(x) = \begin{bmatrix} a & b \\ c & d \end{bmatrix} \in M_2(\mathbf{R})$$

then

$$j(x') = \begin{bmatrix} d & -c \\ -b & a \end{bmatrix}, \quad j(x^*) = \begin{bmatrix} a & c \\ b & d \end{bmatrix}, \quad j(\bar{x}) = \begin{bmatrix} d & -b \\ -c & a \end{bmatrix},$$

and

$$j(x\bar{x}) = j(\bar{x}x) = \det \begin{bmatrix} a & b \\ c & d \end{bmatrix} .I.$$

6.3 Clifford's theorem

In [Cli1], Clifford showed that the even Clifford algebra of a Euclidean space of odd dimension $2k + 1$ can be generated by k commuting four-dimensional algebras, each isomorphic to $\mathbf{H}$ or $M_2(\mathbf{R})$ (the so-called *quaternion algebras*). Since $\mathcal{A}_{2k+1}^+$ is isomorphic to $\mathcal{A}_{2k}$, this result also applies to Clifford algebras of Euclidean spaces of even dimension. It extends easily to Clifford algebras of regular quadratic spaces of even dimension, and can conveniently be expressed in terms of tensor products.

Theorem 6.3.1 *Suppose that (E, q) is a regular quadratic space of even dimension $2k$ and that E is an orthogonal direct sum $F \oplus G$, where F and G are regular subspaces of dimensions $2k - 2$ and 2 respectively. Let ω_F be a volume element in the subalgebra $\mathcal{A}_F = \mathcal{A}(F, q)$ of $\mathcal{A} = \mathcal{A}(E, q)$, and let (g_1, g_2) be an orthogonal basis for G. Let $c_1 = \omega_F g_1$ and $c_2 = \omega_F g_2$, and let C be the subalgebra of $\mathcal{A}$ generated by $\{c_1, c_2\}$. Then C is four-dimensional, and $\mathcal{A}_F$ and C commute. Thus $\mathcal{A} \cong \mathcal{A}_F \otimes C$.*

If G is hyperbolic, or if $g_1^2 = g_2^2 = \omega_F^2$ then $C \cong M_2(\mathbf{R})$. If $g_1^2 = g_2^2 = -\omega_F^2$, then $C \cong \mathbf{H}$.

Proof Since $\dim F$ is even, $c_i = g_i \omega_F$ and $c_i x = x c_i$ for $x \in F$, for $i = 1, 2$. Thus $\mathcal{A}_F$ and C commute, so that $\mathcal{A} \cong \mathcal{A}_F \otimes C$. Also $c_i^2 = \omega_F^2 g_i^2 = \pm 1$, for $i = 1, 2$, and

$$c_1 c_2 = \omega_F g_1 \omega_F g_2 = \omega_F^2 g_1 g_2 = -\omega_F^2 g_2 g_1 = -\omega_F g_2 \omega_F g_1 = -c_2 c_1,$$

so that $(c_1 c_2)^2 = (g_1 g_2)^2 = -g_1^2 g_2^2 = \pm 1$. Thus C is four-dimensional.

If G is hyperbolic, then $c_1^2 = -c_2^2$ and $(c_1c_2)^2 = 1$, so that $C \cong M_2(\mathbf{R})$.

If $g_1^2 = g_2^2 = \omega_F^2$, then $c_1^2 = c_2^2 = 1$, and $(c_1c_2)^2 = -1$, so that $C \cong M_2(\mathbf{R})$.

If $g_1^2 = g_2^2 = -\omega_F^2$, then $c_1^2 = c_2^2 = (c_1c_2)^2 = -1$, so that $C \cong \mathbf{H}$. ☐

Corollary 6.3.1 *If $p - m = 2$ or 4 (mod 8) then $\mathcal{A}_{p,m} \cong M_{2^{k-1}}(\mathbf{H})$ and if $p - m = 0$ or 6 (mod 8) then $\mathcal{A}_{p,m} \cong M_{2^k}(\mathbf{R})$*

Proof By the theorem, $\mathcal{A}_{p+1,m+1} \cong \mathcal{A}_{p,m} \otimes M_2(\mathbf{R}) \cong M_2(\mathcal{A}_{p,m})$, and so $\mathcal{A}_{p+j,m+j} \cong M_{2^j}(\mathcal{A}_{p,m})$. It is therefore sufficient to prove the result when $m = 0$ or $p = 0$.

Suppose that $m = 0$. We prove the result by induction. Suppose that $\mathcal{A}_{8j} \cong M_{2^{4j}}(\mathbf{R})$. Then

$$\mathcal{A}_{8j+2} \cong M_{2^{4j}}(\mathbf{R}) \otimes \mathbf{H} \cong M_{2^{4j}}(\mathbf{H}),$$
$$\mathcal{A}_{8j+4} \cong M_{2^{4j}}(\mathbf{H}) \otimes M_2(\mathbf{R}) \cong M_{2^{4j+1}}(\mathbf{H}),$$
$$\mathcal{A}_{8j+6} \cong M_{2^{4j+1}}(\mathbf{H}) \otimes \mathbf{H}) \cong M_{2^{4j+1}}(\mathbf{R}) \otimes \mathbf{H} \otimes \mathbf{H} \cong M_{2^{4j+3}}(\mathbf{R}),$$
$$\mathcal{A}_{8j+8} \cong M_{2^{4j+3}}(\mathbf{R}) \otimes M_2(\mathbf{R}) \cong M_{2^{4j+4}}(\mathbf{R}).$$

A similar proof establishes the result when $p = 0$. ☐

Proposition 6.3.1 *If $d = p + m = 2k + 1$ and $p - m = 1$ (mod 4) then $\mathcal{A}_{p,m} \cong M_{2^k}(\mathbf{C})$.*

Proof Let F be a regular subspace of $\mathbf{R}_{p,m}$ of dimension $2k$, and let $\mathcal{A}_F = \mathcal{A}(F, q)$. If e_Ω is the volume element of $\mathcal{A}_{p,m}$ then $C = \text{span}(1, e_\Omega)$ is a subalgebra of $\mathcal{A}_{p,m}$ isomorphic to $\mathbf{C}$, and the subalgebras $\mathcal{A}_F$ and C commute. Since $\mathcal{A}_F$ and C generate $\mathcal{A}_{p,m}$ and $\dim \mathcal{A}_{p,m} = (\dim \mathcal{A}_F)(\dim C)$, it follows that $\mathcal{A}_{p,m} \cong \mathcal{A}_F \otimes C \cong \mathcal{A}_F \otimes \mathbf{C}$. If $\mathcal{A}_F \cong M_{2^k}(\mathbf{R})$ then clearly $\mathcal{A}_{p,m} \cong M_{2^k}(\mathbf{C})$. If $\mathcal{A}_F \cong M_{2^{k-1}}(\mathbf{H})$ then

$$\mathcal{A}_{p,m} \cong M_{2^{k-1}}(\mathbf{R}) \otimes \mathbf{H} \otimes \mathbf{C} \cong M_{2^{k-1}}(\mathbf{R}) \otimes M_2(\mathbf{C}) \cong M_{2^k}(\mathbf{C}).$$

☐

We shall consider the case where $p - m = 3$ (mod 4) in the next section.

6.4 Classifying even Clifford algebras

Even Clifford algebras are isomorphic to universal Clifford algebras.

Theorem 6.4.1 $\mathcal{A}_{p+1,m}^+ \cong \mathcal{A}_{p,m}$ and $\mathcal{A}_{p,m+1}^+ \cong \mathcal{A}_{m,p}$.

Proof Let $(e_1, \ldots, e_{p+m+1})$ be the standard basis for $\mathbf{R}_{p+1,m}$ and let $(f_1, \ldots, f_{p+m})$ be the standard basis for $\mathbf{R}_{p,m}$. We consider $\mathbf{R}_{p+1,m}$ as a linear subspace of $\mathcal{A}_{p+1,m}$. Since $e_{i+1}e_{j+1} = (e_1 e_{i+1})(e_1 e_{j+1})$, the set $\{e_1 e_{j+1} : 1 \leq j \leq p+m\}$ generates $\mathcal{A}_{p+1,m}^+$. Since

$$(e_1 e_{j+1})^2 = e_{j+1}^2 = -q(e_{j+1}) = -q(f_j),$$

the mapping $f_j \to e_1 e_{j+1} : 1 \leq j \leq p+m$ from $\mathbf{R}_{p,m}$ into $\mathcal{A}_{p+1,m}^+$ extends by linearity to a Clifford mapping l_1 from $\mathbf{R}_{p,m}$ to $\mathcal{A}_{p+1,m}^+$; this then extends to an algebra homomorphism of $\mathcal{A}_{p,m}$ onto $\mathcal{A}_{p+1,m}^+$. Since $\mathcal{A}_{p,m}$ and $\mathcal{A}_{p+1,m}^+$ both have dimension 2^{p+m}, this homomorphism is an isomorphism.

Similarly, let $(e_1, \ldots, e_{p+m+1})$ be the standard basis for $\mathbf{R}_{p,m+1}$, and let $(g_1, \ldots, g_{p+m})$ be the standard basis for $\mathbf{R}_{m,p}$. Let $d = p + m$. We consider $\mathbf{R}_{p,m+1}$ as a linear subspace of $\mathcal{A}_{p,m+1}$. As above, the set $\{e_j e_{d+1} : 1 \leq j \leq d\}$ generates $\mathcal{A}_{p,m+1}^+$. Since $(e_j e_{d+1})^2 = -e_j^2 = q(e_j)$, the mapping $g_j \to e_{d+1-j}e_{d+1} : 1 \leq j \leq d$ from $\mathbf{R}_{m,p}$ to $\mathcal{A}_{p,m+1}^+$ extends by linearity to a Clifford mapping r_{d+1} from $\mathbf{R}_{m,p}$ to $\mathcal{A}_{p,m+1}^+$; this then extends to an algebra homomorphism of $\mathcal{A}_{m,p}$ onto $\mathcal{A}_{p+1,m}^+$. Again, $\mathcal{A}_{p,m}$ and $\mathcal{A}_{p+1,m}^+$ both have dimension 2^{p+m}, so that this homomorphism is an isomorphism. $\qquad\square$

Corollary 6.4.1 *Suppose that $d = 2k + 1$ is odd.*
(i) If $p - m = 3$ (mod 8) then $A_{p,m} \cong M_{2^{k-1}}(\mathbf{H}) \oplus M_{2^{k-1}}(\mathbf{H})$.
(ii) If $p - m = 7$ (mod 8) then $A_{p,m} \cong M_{2^k}(\mathbf{R}) \oplus M_{2^k}(\mathbf{R})$.

Proof For $\mathcal{B}_{p,m} \cong \mathcal{A}_{p,m}^+ \cong \mathcal{A}_{p-1,m}$ if $p > 0$ and $\mathcal{B}_{0,m} \cong \mathcal{A}_{0,m}^+ \cong \mathcal{A}_{m-1}$ if $p = 0$. $\qquad\square$

6.5 Cartan's periodicity law

It follows from Theorem 6.4.1 that $\mathcal{A}_{p,m+1} \cong \mathcal{A}_{m,p+1}$, since each is isomorphic to $\mathcal{A}_{p+1,m+1}^+$.

This result needs treating with caution, since in general $\mathcal{A}_{p,m+1}$ is not isomorphic as a super-algebra to $\mathcal{A}_{m,p+1}$. For $\mathcal{A}_{p,m+1}^+ \cong \mathcal{A}_{m,p}$ and $\mathcal{A}_{p+1,m}^+ \cong \mathcal{A}_{p,m}$, and in general the algebras $\mathcal{A}_{m,p}$ and $\mathcal{A}_{p,m}$ are not isomorphic. For example, if $p - m = 3 \,(\text{mod}\,4)$ then $\mathbf{Z}(\mathcal{A}_{p,m}) \cong \mathbf{R}$ and $\mathbf{Z}(\mathcal{A}_{m,p}) \cong \mathbf{C}$.

We obtain further results when p or m is at least 4.

Proposition 6.5.1 *There is a graded isomorphism of $\mathcal{A}_{p,m+4}$ onto $\mathcal{A}_{p+4,m}$.*

Let us denote the standard basis of $\mathbf{R}_{p+4,m}$ by $(e_1, \ldots, e_d)$ and the standard basis of $\mathbf{R}_{p,m+4}$ by $(g_1, \ldots, g_d)$, where $d = p+4+m$. As usual, we consider $\mathbf{R}_{p+4,m}$ as a linear subspace of $\mathcal{A}_{p+4,m}$, and $\mathbf{R}_{p,m+4}$ as a linear subspace of $\mathcal{A}_{p,m+4}$. Let $f = e_{p+1}e_{p+2}e_{p+3}e_{p+4} \in \mathcal{A}_{p+4,m}$, and let $f_j = e_j f$ for $1 \leq j \leq d$. Then $f^2 = 1$, and

$$f_j^2 = e_j^2 = -1 \quad \text{for } 1 \leq j \leq p,$$
$$f_j^2 = -e_j^2 = 1 \quad \text{for } p+1 \leq j \leq p+4, \text{ and}$$
$$f_j^2 = e_j^2 = 1 \quad \text{for } p+5 \leq j \leq d.$$

Also

$$f_j f_k = -f_k f_j \text{ for } j \neq k.$$

Thus the mapping π from $\mathbf{R}_{p,m+4}$ to $\mathcal{A}_{p+4,m}$ defined by $\pi(\sum_{j=1}^{d} x_j g_j) = \sum_{j=1}^{d} x_j f_j$ is a Clifford mapping. This extends to an isomorphism π of $\mathcal{A}_{p,m+4}$ onto $\mathcal{A}_{p+4,m}$. Since $f_j \in \mathcal{A}_{p+4,m}^-$ for $1 \leq j \leq d$, $\pi(\mathcal{A}_{p,m+4}^-) = \mathcal{A}_{p+4,m}^-$ and $\pi(\mathcal{A}_{p,m+4}^+) = \mathcal{A}_{p+4,m}^+$.

We have the following consequence.

Theorem 6.5.1 (Cartan's periodicity law) *There are graded isomorphisms between the three graded algebras $\mathcal{A}_{p+8,m}$, $\mathcal{A}_{p,m+8}$ and $M_{16}(\mathcal{A}_{p,m})$.*

Proof For

$$\mathcal{A}_{p+8,m} \cong \mathcal{A}_{p+4,m+4} \cong \mathcal{A}_{p,m+8}, \text{ and } \mathcal{A}_{p+4,m+4} \cong M_{16}(\mathcal{A}_{p,m}),$$

and the isomorphisms respect the gradings. □

It is therefore sufficient to tabulate the results for $0 \leq, p, m \leq 7$. We have the following table (where K^2 denotes $K \oplus K$). Note that these results are obtained without appealing to Wedderburn's theorem. Further, Theorem 6.4.1 allows us to classify the corresponding even Clifford algebras.

Universal Clifford algebras

m \ p	0	1	2	3	4	5	6	7
0	$\mathbf{R}$	$\mathbf{C}$	$\mathbf{H}$	$\mathbf{H}^2$	$M_2(\mathbf{H})$	$M_4(\mathbf{C})$	$M_8(\mathbf{R})$	$M_8(\mathbf{R}^2)$
1	$\mathbf{R}^2$	$M_2(\mathbf{R})$	$M_2(\mathbf{C})$	$M_2(\mathbf{H})$	$M_2(\mathbf{H}^2)$	$M_4(\mathbf{H})$	$M_8(\mathbf{C})$	$M_8(\mathbf{R})$
2	$M_2(\mathbf{R})$	$M_2(\mathbf{R}^2)$	$M_4(\mathbf{R})$	$M_4(\mathbf{C})$	$M_4(\mathbf{H})$	$M_4(\mathbf{H}^2)$	$M_8(\mathbf{H})$	$M_{16}(\mathbf{C})$
3	$M_2(\mathbf{C})$	$M_4(\mathbf{R})$	$M_4(\mathbf{R}^2)$	$M_8(\mathbf{R})$	$M_8(\mathbf{C})$	$M_8(\mathbf{H})$	$M_8(\mathbf{H}^2)$	$M_{16}(\mathbf{H})$
4	$M_2(\mathbf{H})$	$M_4(\mathbf{C})$	$M_8(\mathbf{R})$	$M_8(\mathbf{R}^2)$	$M_{16}(\mathbf{R})$	$M_{16}(\mathbf{C})$	$M_{16}(\mathbf{H})$	$M_{16}(\mathbf{H}^2)$
5	$M_2(\mathbf{H}^2)$	$M_4(\mathbf{H})$	$M_8(\mathbf{C})$	$M_{16}(\mathbf{R})$	$M_{16}(\mathbf{R}^2)$	$M_{32}(\mathbf{R})$	$M_{32}(\mathbf{C})$	$M_{32}(\mathbf{H})$
6	$M_4(\mathbf{H})$	$M_4(\mathbf{H}^2)$	$M_8(\mathbf{H})$	$M_{16}(\mathbf{C})$	$M_{32}(\mathbf{R})$	$M_{32}(\mathbf{R}^2)$	$M_{64}(\mathbf{R})$	$M_{64}(\mathbf{C})$
7	$M_8(\mathbf{C})$	$M_8(\mathbf{H})$	$M_8(\mathbf{H}^2)$	$M_{16}(\mathbf{H})$	$M_{32}(\mathbf{C})$	$M_{64}(\mathbf{R})$	$M_{64}(\mathbf{R}^2)$	$M_{128}(\mathbf{R})$

6.6 Classifying complex Clifford algebras

Let us briefly describe what happens in the complex case. This is much less interesting than the real case. Suppose that (E,q) is a regular complex quadratic space, with standard orthogonal basis $(e_1,\ldots,e_d)$. We leave the reader to verify the following results. The notions of Clifford mapping, Clifford algebra and universal Clifford algebra $\mathcal{C}(E,q)$ are defined as in the real case. The regular quadratic space (E,q) has a universal Clifford algebra (this is most easily seen from the constructions that we shall make), and for such algebras we can define the *principal involution, reversal* and *conjugation* (this conjugation, which is a linear anti-involution, is quite different from the conjugation of complex numbers, which is anti-linear, mapping z to $\bar z$). We can therefore define the even Clifford algebra $\mathcal{C}(E,q)^+$. Then $\mathcal{C}(E,q)^+$ is isomorphic to $\mathcal{C}(F,q)$, where $F = \text{span}(e_2,\ldots,e_d)$.

If $\dim E = 1$ then the mapping $\lambda e_1 \to (i\lambda, -i\lambda)$ is a Clifford mapping of (E,q) into $\mathbf{C}^2$, and the resulting Clifford algebra is isomorphic to the commutative complex algebra $\mathbf{C}^2$, and is not simple.

If $\dim E = 2$ then the mapping

$$\lambda_1 e_1 + \lambda_2 e_2 \to \lambda_1 J + i\lambda_2 Q = \begin{bmatrix} 0 & -\lambda_1 - i\lambda_2 \\ \lambda_1 - i\lambda_2 & 0 \end{bmatrix}$$

is a Clifford mapping of (E,q) into $M_2(\mathbf{C})$, and the resulting Clifford algebra is isomorphic to the algebra $M_2(\mathbf{C})$, and is simple.

If $\dim E > 2$ then we can write $E = \text{span}(e_1,e_2) \oplus G$, where $G = \text{span}(e_3,\ldots,e_d)$. The mapping

$$\lambda_1 e_1 + \lambda_2 + g \to \lambda_1 I \otimes J + i\lambda_2 I \otimes Q + iy \otimes U = \begin{bmatrix} iy & -\lambda_1 - i\lambda_2 \\ \lambda_1 - i\lambda_2 & -iy \end{bmatrix}$$

is a Clifford mapping of (E,q) into $M_2(\mathcal{C}(G,q))$, and the resulting Clifford algebra is isomorphic to the algebra $M_2(\mathcal{C}(G,q))$, and is simple if and only if $\mathcal{C}(G,q)$ is. As a consequence, we have the following theorem.

Theorem 6.6.1 *Suppose that (E,q) is a regular complex quadratic space. If $\dim E = 2k$ is even then the universal Clifford algebra $\mathcal{C}(E,q)$ is isomorphic to $M_{2^k}(\mathbf{C})$, and is simple. If $\dim E = 2k+1$ is odd then $\mathcal{C}(E,q)$ is isomorphic to $M_{2^k}(\mathbf{C}) \oplus M_{2^k}(\mathbf{C})$, and is not simple. The even Clifford algebra $\mathcal{C}(E,q)^+$ is isomorphic to $\mathcal{C}(F,q')$, where (F,q') is a regular complex quadratic space with dimension one less than the dimension of E.*

7

Representing Clifford algebras

We have seen that the Clifford algebras and the even Clifford algebras of regular quadratic spaces are isomorphic either to a full matrix algebra $M_k(\mathbf{D})$ or to a direct sum $M_k(\mathbf{D}) \oplus M_k(\mathbf{D})$ of two full matrix algebras, where $\mathbf{D} = \mathbf{R}, \mathbf{C}$ or $\mathbf{H}$. These algebras act naturally on a real or complex vector space, or on a left $\mathbf{H}$-module, respectively. In this way, we obtain representations of the Clifford algebras.

In this chapter, we shall construct some of these representations, using tensor products. These representations give useful information about the algebra, and its relation to its even subalgebra. We then construct some explicit representations of low-dimensional Clifford algebras. These are useful in practice, but it is probably not necessary to consider each of them in detail, on the first reading.

7.1 Spinors

When $\mathcal{A}_{p,m}$ is simple, we have represented it as $M_k(\mathbf{D})$, where $\mathbf{D} = \mathbf{R}, \mathbf{C}$ or $\mathbf{H}$, so that we can consider $\mathcal{A}_{p,m}$ acting on the left $\mathbf{D}$-module $\mathbf{D}^k$. A left $\mathbf{R}$-module is just a real vector space, and a left $\mathbf{C}$-module is a complex vector space. Since $\mathbf{H}$ is a division algebra, the notions of linear independence, basis, and dimension can be defined as easily for a left $\mathbf{H}$-module as for a vector space, and a left $\mathbf{H}$-module is frequently called a *vector space over* $\mathbf{H}$. When $\mathcal{A}_{p,m}$ is not simple, then $\mathcal{A}_{p,m} \cong M_k(\mathbf{D}) \oplus M_k(\mathbf{D})$, and so we can consider $\mathcal{A}_{p,m}$ acting on $\mathbf{D}^k \oplus \mathbf{D}^k$. It follows from Theorem 2.7.3 that there are no essentially different representations.

If $A_{p,m}$ is simple and π is an irreducible representation of $A_{p,m}$ in $L(W)$ then W is called a *spinor* space, and the elements of W are called spinors. For example if we consider the left regular representation of $A_{p,m}$ then the minimal left ideals of $A_{p,m}$ are spinor spaces; if we identify $A_{p,m}$ with $M_k(\mathbf{D})$ then the columns of $M_k(\mathbf{D})$ are spinor spaces. It follows from Theorem 2.7.3 that any two irreducible representations are equivalent. We denote a spinor space for $A_{p,m}$ by $S_{p,m}$, and consider it as a left $A_{p,m}$-module.

If $A_{p,m}$ is simple and $A_{p,m} \cong M_k(\mathbf{R})$ then

$$\dim A_{p,m} = 2^d = \dim M_k(\mathbf{R}) = k^2,$$

so that $d = 2t$ is even, and $k = 2^t$. Thus the real dimension of a spinor space is $2^{d/2}$.

If $A_{p,m}$ is simple and $A_{p,m} \cong M_k(\mathbf{C})$ then

$$\dim_{\mathbf{R}} A_{p,m} = 2^d = \dim_{\mathbf{R}} M_k(\mathbf{C}) = 2k^2,$$

so that $d = 2t + 1$ is odd, and $k = 2^t$. Thus the complex dimension of a spinor space is $2^{(d-1)/2}$, and its real dimension is $2^{(d+1)/2}$.

If $A_{p,m}$ is simple and $A_{p,m} \cong M_k(\mathbf{H})$ then

$$\dim_{\mathbf{R}} A_{p,m} = 2^d = \dim_{\mathbf{R}} M_k(\mathbf{C}) = 4k^2,$$

so that $d = 2t$ is even, and $k = 2^{t-1}$. Thus the quaternionic dimension of a spinor space is $2^{(d-2)/2}$, and its real dimension is $2^{(d+2)/2}$.

When $p - m = 3 \pmod 4$, so that $A_{p,m}$ is not simple, we consider irreducible representations of the simple non-universal Clifford algebra $B_{p,m}$; the resulting left $B_{p,m}$-modules are called *semi-spinor* spaces.

We then have the table on the following page.

Suppose that (E, q) is a Euclidean space. If W is a spinor space for (E, q), we can consider W as a minimal left ideal in $A(E, q)$, and give W the inner product and norm inherited from $(A(E, q), Q)$. If $a \in N_1(E, q)$ then the mapping $w \to aw : W \to W$ is an isometry of W.

Things are quite different if q is not positive definite. Suppose that $q(x) = -1$. Let $p = \frac{1}{2}(1 + x)$. Then $p^2 = p$, $\bar{p} = 1 - p$ and $p\bar{p} = 0$. Thus $A(E, q) = A(E, q)p \oplus A(E, q)(1 - p)$ is the vector space direct sum of two left ideals. If $ap \in A(E, q)p$ then

$$Q(ap) = \tau_n(ap\bar{p}\bar{a}) = 0,$$

and similarly Q vanishes on $A(E, q)\bar{p}$. If W is a minimal left ideal of $A(E, q)$ then W is contained in one of $A(E, q)p$ or $A(E, q)\bar{p}$, and so W is an isotropic subspace of $(A(E, q), Q)$.

Spinor and [semi-spinor] spaces

m		$p \rightarrow$							
		0	1	2	3	4	5	6	7
	0	R	C	H	$[H]$	H^2	C^4	R^8	$[R^8]$
$\downarrow$	1	$[R]$	R^2	C^2	H^2	$[H^2]$	H^4	C^8	R^{16}
	2	R^2	$[R^2]$	R^4	C^4	H^4	$[H^4]$	H^8	C^{16}
	3	C^2	R^4	$[R^4]$	R^8	C^8	H^8	$[H^8]$	H^{16}
	4	H^2	C^4	R^8	$[R^8]$	R^{16}	C^{16}	H^{16}	$[H^{16}]$
	5	$[H^2]$	H^4	C^8	R^{16}	$[R^{16}]$	R^{32}	C^{32}	H^{32}
	6	H^4	$[H^4]$	H^8	C^{16}	R^{32}	$[R^{32}]$	R^{64}	C^{64}
	7	C^8	H^8	$[H^8]$	H^{16}	C^{32}	R^{64}	$[R^{64}]$	R^{128}

7.2 The Clifford algebras $\mathcal{A}_{k,k}$

Recall that $\mathcal{A}_{k,k} \cong M_{2^k}(\mathbf{R})$, which we can identify with the k-fold tensor product $\otimes^k M_2(\mathbf{R})$ of four-dimensional algebras. Since we shall consider different values of k, we shall denote the standard orthogonal basis of $\mathbf{R}_{k,k}$ by $(e_1^{(k)}, \dots, e_k^{(k)}, f_1^{(k)}, \dots, f_k^{(k)})$. Recall the definitions of the matrices I, J, Q and U made in Section 2.2. Let us set

$$j_k(e_i^{(k)}) = (\otimes^{k-i} I) \otimes J \otimes (\otimes^{i-1} Q) \text{ for } 1 \le i \le k,$$
$$j_k(f_1^{(k)}) = \otimes^k Q,$$
$$j_k(f_i^{(k)}) = (\otimes^{i-2} I) \otimes U \otimes (\otimes^{k+1-i} Q) \text{ for } 2 \le i \le k.$$

For example, when $k = 4$,

$$j_4(e_1^{(4)}) = I \otimes I \otimes I \otimes J,$$
$$j_4(e_2^{(4)}) = I \otimes I \otimes J \otimes Q,$$
$$j_4(e_3^{(4)}) = I \otimes J \otimes Q \otimes Q,$$
$$j_4(e_4^{(4)}) = J \otimes Q \otimes Q \otimes Q,$$
$$j_4(f_1^{(4)}) = Q \otimes Q \otimes Q \otimes Q,$$
$$j_4(f_2^{(4)}) = U \otimes Q \otimes Q \otimes Q,$$
$$j_4(f_3^{(4)}) = I \otimes U \otimes Q \otimes Q,$$
$$j_4(f_4^{(4)}) = I \otimes I \otimes U \otimes Q.$$

Then the conditions of Theorem 5.1.1 are satisfied, and so j_k extends to a Clifford mapping of $\mathbf{R}_{k,k}$ into $M_{2^k}(\mathbf{R})$, and this in turn extends to an algebra isomorphism of $\mathcal{A}_{k,k}$ onto $M_{2^k}(\mathbf{R})$.

If we write $M_{2^k}(\mathbf{R})$ as the tensor product $M_{2^{k-1}}(\mathbf{R}) \otimes M_2(\mathbf{R})$, and if $x \in \mathbf{R}_{k,k}$, then the non-zero elements of $j_k(x)$ occur only in the top right-hand and bottom left-hand quadrants. It therefore follows that that if $a \in \mathcal{A}_{k,k}^+$, then the non-zero elements of $j_k(a)$ occur only in the top left-hand and bottom right-hand quadrants, and if $a \in \mathcal{A}_{k,k}^-$, then the non-zero elements of $j_k(a)$ occur only in the top right-hand and bottom left-hand quadrants. Schematically we have:

$$\begin{bmatrix} [\mathcal{A}^+] & [\mathcal{A}^-] \\ [\mathcal{A}^-] & [\mathcal{A}^+] \end{bmatrix}.$$

We denote the volume element of $\mathcal{A}_{k,k}$ by $V_{k,k}$: easy calculations show that $j_k(V_{k,k}) = \pm(\otimes^{k-1} I) \otimes U$.

We now consider $\mathcal{A}_{k+1,k+1}^+$. Let

$$g_i^{(k+1)} = e_1^{(k+1)} e_{i+1}^{(k+1)} \quad \text{for } 1 \leq i \leq k,$$
$$h_i^{(k+1)} = e_1^{(k+1)} f_{i+1}^{(k+1)} \quad \text{for } 1 \leq i \leq k+1.$$

The elements $g_i^{(k+1)}$ and $h_i^{(k+1)}$ are in $\mathcal{A}_{k+1,k+1}^+$. Then more easy calculations show that

$$j_{k+1}(g_i^{(k+1)}) = j_k(e_i^{(k)}) \otimes U \quad \text{and} \quad j_{k+1}(h_i^{(k+1)}) = j_k(f_i^{(k)}) \otimes U,$$

for $1 \leq i \leq k$. For example, when $k = 3$ then

$$j_4(g_1^{(4)}) = (I \otimes I \otimes J) \otimes U,$$
$$j_4(g_2^{(4)}) = (I \otimes J \otimes Q) \otimes U,$$
$$j_4(g_3^{(4)}) = (J \otimes Q \otimes Q) \otimes U,$$
$$j_4(h_1^{(4)}) = (Q \otimes Q \otimes Q) \otimes U,$$
$$j_4(h_2^{(4)}) = (U \otimes Q \otimes Q) \otimes U,$$
$$j_4(h_3^{(4)}) = (I \otimes U \otimes Q) \otimes U,$$
$$j_4(h_4^{(4)}) = (I \otimes I \otimes U) \otimes U.$$

Let us define $\phi_k(e_i^{(k)}) = j_{k+1}(g_i^{(k+1)})$ and $\phi_k(f_i^{(k)}) = j_{k+1}(h_i^{(k+1)})$ for $1 \leq i \leq k$, and extend by linearity. Then ϕ_k is a Clifford mapping of $\mathbf{R}_{k,k}$ into $j_{k+1}(\mathcal{A}_{k+1,k+1}^+)$, and therefore extends to an algebra isomorphism of $\mathcal{A}_{k,k}$ into $j_{k+1}(\mathcal{A}_{k+1,k+1}^+)$. Then

$$\phi_k(a) = \begin{bmatrix} j_k(a) & 0 \\ 0 & j_k(a') \end{bmatrix}.$$

Thus the top left-hand quadrant of $\phi_k(a)$ is exactly the same as the matrix $j_k(a)$.

The algebra $j_{k+1}(\mathcal{A}_{k+1,k+1}^+)$ is then generated by $\phi_k(\mathcal{A}_{k,k})$ and $j_{k+1}(V_{k+1,k+1}) = \pm(\otimes^k I) \otimes U$.

Exercise

1. Prove directly by induction that $j_k(\mathbf{R}_{k,k})$ generates $M_{2^k}(\mathbf{R})$. Deduce another proof of the existence of the universal Clifford algebra $\mathcal{A}_{k,k}$. Use this to prove the existence of a universal Clifford algebra $\mathcal{A}(E, q)$ for a regular quadratic space (E, q).

7.3 The algebras $\mathcal{B}_{k,k+1}$ and $\mathcal{A}_{k,k+1}$

We can use the ideas of the preceding section to construct representations of the non-universal Clifford algebra $\mathcal{B}_{k,k+1}$ and the universal Clifford algebra $\mathcal{A}_{k,k+1}$. Let $(e_1^{(k)}, \ldots, e_k^{(k)}, f_1^{(k)}, \ldots, f_{k+1}^{(k)})$ denote the standard orthogonal basis of $\mathbf{R}_{k,k+1}$.

Let us set

$$l_k(e_i^{(k)}) = (\otimes^{k-i} I) \otimes J \otimes (\otimes^{i-1} Q) \quad \text{for } 1 \leq i \leq k,$$
$$l_k(f_1^{(k)}) = \otimes^k Q,$$
$$l_k(f_i^{(k)}) = (\otimes^{i-2} I) \otimes U \otimes (\otimes^{k+1-i} Q) \quad \text{for } 2 \leq i \leq k+1,$$

and extend by linearity to obtain a linear mapping l_k from $\mathbf{R}_{k,k+1}$ into $M_{2^k}(\mathbf{R})$.

For example, when $k = 3$,

$$l_4(e_1^{(3)}) = I \otimes I \otimes J,$$
$$l_4(e_2^{(3)}) = I \otimes J \otimes Q,$$
$$l_4(e_3^{(3)}) = J \otimes Q \otimes Q,$$
$$l_4(f_1^{(3)}) = Q \otimes Q \otimes Q,$$
$$l_4(f_2^{(4)}) = U \otimes Q \otimes Q,$$
$$l_4(f_3^{(4)}) = I \otimes U \otimes Q,$$
$$l_4(f_4^{(4)}) = I \otimes I \otimes U.$$

Then l_k is a Clifford mapping, and so extends to an algebra homomorphism of $\mathcal{A}_{k,k+1}$ into $M_{2^k}(\mathbf{R})$. Since $l_k(e_\Omega) = -\otimes^k I$, l_k is not injective, and so, since $\mathcal{B}_{k,k+1} \cong \mathcal{A}_{k,k+1}/\text{span}(I, e_\Omega)$, we obtain an isomorphic representation of $\mathcal{B}_{k,k+1}$ onto $M_{2^k}(\mathbf{R})$.

We can easily adjust this to obtain a representation of $\mathcal{A}_{k,k+1}$. Let us set

$$m_k(e_i^{(k)}) = l_k(e_i^{(k)}) \otimes U \quad \text{for } 1 \leq i \leq k,$$
$$m_k(f_i^{(k)}) = l_k(f_i^{(k)}) \otimes U \quad \text{for } 1 \leq i \leq k+1,$$

and extend by linearity to obtain a linear mapping l_k from $\mathbf{R}_{k,k+1}$ into $M_{2^{k+1}}(\mathbf{R})$.

For example, when $k = 3$,

$$m_4(e_1^{(3)}) = I \otimes I \otimes J \otimes U,$$
$$m_4(e_2^{(3)}) = I \otimes J \otimes Q \otimes U,$$
$$m_4(e_3^{(3)}) = J \otimes Q \otimes Q \otimes U,$$
$$m_4(f_1^{(3)}) = Q \otimes Q \otimes Q \otimes U,$$
$$m_4(f_2^{(4)}) = U \otimes Q \otimes Q \otimes U,$$
$$m_4(f_3^{(4)}) = I \otimes U \otimes Q \otimes U,$$
$$m_4(f_4^{(4)}) = I \otimes I \otimes U \otimes U.$$

Then m_k is a Clifford mapping, and so extends to an algebra homomorphism of $\mathcal{A}_{k,k+1}$ into $M_{2^{k+1}}(\mathbf{R})$. Since $m_k(e_\Omega) = -(\otimes^k I) \otimes U$, m_k is injective. Bearing in mind that the last factor is U, it follows that

$$m_k(\mathcal{A}_{k,k+1}) = \left\{ \begin{bmatrix} a & 0 \\ 0 & b \end{bmatrix} : a, b \in M_{2^k}(\mathbf{R}) \right\} \cong M_{2^k}(\mathbf{R}) \oplus M_{2^k}(\mathbf{R}).$$

7.4 The algebras $\mathcal{A}_{k+1,k}$ and $\mathcal{A}_{k+2,k}$

We can also use the ideas of Section 7.2 to obtain a representation of $\mathcal{A}_{k+1,k}$ as $M_{2^k}(\mathbf{C})$ and a representation of $\mathcal{A}_{k+2,k}$ as $M_{2^k}(\mathbf{H})$.

Let $(e_0^{(k)}, \ldots, e_k^{(k)}, f_1^{(k)}, \ldots, f_k^{(k)})$ denote the standard orthogonal basis of $\mathbf{R}_{k+1,k}$. We can consider the mapping j_k of Section 7.2 as a Clifford mapping from $\mathbf{R}_{k,k}$ into $M_{2^k}(\mathbf{C})$. We can extend j_k as a mapping from $\mathbf{R}_{k+1,k}$ into $M_{2^k}(\mathbf{C})$ by setting $j_k(e_0^{(k)}) = -i(\otimes^{k-1}I) \otimes U$, and extending by linearity. Then j_k is a Clifford mapping of $\mathbf{R}_{k+1,k}$ into $M_{2^k}(\mathbf{C})$ which extends to an algebra isomorphism of $\mathcal{A}_{k+1,k}$ onto $M_{2^k}(\mathbf{C})$. In this representation, $j_k(e_\Omega) = iI$.

Similarly, we can take a basis $(d_1^{(k)}, d_2^{(k)}, e_1^{(k)}, \ldots, e_k^{(k)}, f_1^{(k)}, \ldots, f_k^{(k)})$ for $\mathbf{R}_{k+2,k}$ and consider j_k as a Clifford mapping from $\mathbf{R}_{k,k}$ into $M_{2^k}(\mathbf{H})$. We can extend j_k as a mapping from $\mathbf{R}_{k+2,k}$ into $M_{2^k}(\mathbf{H})$ by setting $j_k(d_1^{(k)}) = \mathbf{i}(\otimes^{k-1}I) \otimes U$, $j_k(d_2^{(k)}) = \mathbf{j}(\otimes^{k-1}I) \otimes U$, and extending by linearity. Then j_k is a Clifford mapping of $\mathbf{R}_{k+2,k}$ into $M_{2^k}(\mathbf{H})$ which extends to an algebra isomorphism of $\mathcal{A}_{k+2,k}$ onto $M_{2^k}(\mathbf{H})$. In this representation, $j_k(e_\Omega) = \mathbf{k}I$.

7.5 Clifford algebras $\mathcal{A}(E,q)$ with $\dim E = 3$

We now consider some explicit representations of low-dimensional Clifford algebras.

The algebra $\mathcal{A}_3$
The algebra $\mathcal{A}_3$ is not simple. Using Theorem 6.4.1,

$$\mathcal{A}_3 \cong \mathcal{A}_3^+ \oplus \mathcal{A}_3^+ \cong \mathcal{A}_2 \oplus \mathcal{A}_2 \cong \mathbf{H} \oplus \mathbf{H}.$$

A Clifford mapping k from $\mathbf{R}_3$ to $\mathbf{H}$ is obtained by setting

$$k(x_1 e_1 + x_2 e_2 + x_3 e_3) = x_1 \boldsymbol{i} + x_2 \boldsymbol{j} + x_3 \boldsymbol{k}.$$

This extends to an algebra isomorphism of $\mathcal{B}_3$ onto $\mathbf{H}$. We then obtain a Clifford mapping of $\mathbf{R}_3$ into $\mathbf{H} \oplus \mathbf{H}$ by setting $j(x) = (k(x), -k(x))$; this extends to an algebra isomorphism of $\mathcal{A}_3$ onto $\mathbf{H} \oplus \mathbf{H}$.

The algebra $\mathcal{A}_{2,1}$
The Clifford algebra $\mathcal{A}_{2,1}$ is isomorphic to $M_2(\mathbf{C})$. There are many representations that we can use; we consider two, each of which has interesting properties.
First, let

$$j_1(x_1 e_1 + x_2 e_2 + x_3 e_3) = ix_1 Q - x_2 J + x_3 U = \begin{bmatrix} x_3 & ix_1 + x_2 \\ ix_1 - x_2 & x_3 \end{bmatrix}.$$

This is a Clifford mapping of $\mathbf{R}_{2,1}$ into $M_2(\mathbf{C})$, which extends to an algebra homomorphism of $\mathcal{A}_{2,1}$ into $M_2(\mathbf{C})$. Since $\mathcal{A}_{2,1}$ is simple, and both algebras have dimension 8, it is an isomorphism. Since $j_1(e_1 e_2) = -iU$, $j_2(e_1 e_3) = iJ$ and $j_1(e_2 e_3) = -Q$,

$$j_1(\mathcal{A}_{2,1}^+) = \left\{ \begin{bmatrix} z & w \\ \bar{w} & \bar{z} \end{bmatrix} : z, w \in \mathbf{C} \right\}.$$

In this representation, it is not apparent that $\mathcal{A}_{2,1}^+$ is isomorphic to $M_2(\mathbf{R})$.
The linear subspace $\operatorname{span}(e_1, e_2)$ of $\mathbf{R}_{2,1}$ is isomorphic to $\mathbf{R}_2$. Thus

$$\tilde{\imath}(\mathcal{A}_2) = \operatorname{span}(I, iQ, -J, -iU) = \left\{ \begin{bmatrix} z & w \\ -\bar{w} & \bar{z} \end{bmatrix} : z, w \in \mathbf{C} \right\} = H,$$

where H is the real subalgebra of $M_2(\mathbf{C})$ which was used to define the quaternions in Section 2.3.

The linear subspace $\text{span}(e_2, e_3)$ of $\mathbf{R}_{2,1}$ is isomorphic to $\mathbf{R}_{1,1}$. Thus

$$\tilde{\imath}(\mathcal{A}_{1,1}) = \text{span}(I, -J, U, Q) = M_2(\mathbf{R}).$$

Secondly, let

$$j_2(x_1 e_1 + x_2 e_2 + x_3 e_3) = i x_1 Q + i x_2 U - i x_3 J = i \begin{bmatrix} x_2 & x_1 + x_3 \\ x_1 - x_3 & -x_2 \end{bmatrix}.$$

Then j_2 is a Clifford mapping of $\mathbf{R}_{2,1}$ into $M_2(\mathbf{C})$, which extends to an algebra homomorphism $j_2 : \mathcal{A}_{2,1} \to M_2(\mathbf{C})$; as before, this is an isomorphism. Note that $j_2(\mathbf{R}_{2,1}) = i M_2^{(0)}(\mathbf{R})$, where

$$M_2^{(0)}(\mathbf{R}) = \{T \in M_2(\mathbf{R}) : \tau(T) = 0\}.$$

In this representation,

$$j_2(e_1 e_2) = -J, \ \ j_2(e_1 e_3) = U, \ \ j_2(e_2 e_3) = -Q,$$

so that

$$j_2(\mathcal{A}_{2,1}^+) = \text{span}(I, -J, U, -Q) = M_2(\mathbf{R}).$$

Once again,

$$\tilde{\imath}(\mathcal{A}_2) = \left\{ \begin{bmatrix} z & w \\ -\bar{w} & \bar{z} \end{bmatrix} : z, w \in \mathbf{C} \right\} = H,$$

while

$$\tilde{\imath}(\mathcal{A}_{1,1}) = \left\{ \begin{bmatrix} z & w \\ \bar{w} & \bar{z} \end{bmatrix} : z, w \in \mathbf{C} \right\}.$$

Again, in this representation, it is not apparent that $\mathcal{A}_{1,1}$ is isomorphic to $M_2(\mathbf{R})$.

The algebra $\mathcal{A}_{1,2}$

The algebra $\mathcal{A}_{1,2}$ is not simple, and

$$\mathcal{A}_{1,2} \cong A_{0,2} \oplus A_{0,2} \cong M_2(\mathbf{R}) \oplus M_2(\mathbf{R}).$$

A Clifford mapping $k : \mathbf{R}_{1,2} \to M_2(\mathbf{R})$ is given by setting

$$k(x_1 e_1 + x_2 e_2 + x_3 e_3) = x_1 J + x_2 Q + x_3 U = \begin{bmatrix} x_3 & -x_1 + x_2 \\ x_1 + x_2 & -x_3 \end{bmatrix},$$

and this extends to an algebra homomorphism of $\mathcal{A}_{1,2}$ onto $M_2(\mathbf{R})$. Then $k(e_\Omega) = -I$, and we have represented $\mathcal{B}_{1,2}$ as $M_2(\mathbf{R})$. Note that $k(\mathbf{R}_{1,2}) = M_2^{(0)}(\mathbf{R}) = \{T \in M_2(\mathbf{R}) : \tau(T) = 0\}.$

If we set $j(x) = (k(x), -k(x))$, we obtain a Clifford mapping of $\mathcal{R}_{1,2}$ into $M_2(\mathbf{R}) \oplus M_2(\mathbf{R})$ which extends to an algebra isomorphism of $\mathcal{A}_{1,2}$ onto $M_2(\mathbf{R}) \oplus M_2(\mathbf{R})$.

The algebra $\mathcal{A}_{0,3}$

The Clifford algebra $\mathcal{A}_{0,3}$ is isomorphic to $M_2(\mathbf{C})$. We use the Pauli spin matrices to obtain an explicit representation. Let

$$j((x_1, x_2, x_3)) = x_1\sigma_1 + x_2\sigma_2 + x_3\sigma_3 = x_1 Q + ix_2 J + x_3 U$$

$$= \begin{bmatrix} x_3 & x_1 - ix_2 \\ x_1 + ix_2 & -x_3 \end{bmatrix}.$$

This is a Clifford mapping of $\mathbf{R}_{0,3}$ into $M_2(\mathbf{C})$, which extends to an isomorphism of $\mathcal{A}_{0,3}$ onto $M_2(\mathbf{C})$.

The matrices σ_1, σ_2 and σ_3 play a symmetric role; yet σ_1 and σ_3 appear to be real, and σ_2 to be complex. Why is this? It follows from Proposition 5.4.1 that the centre $Z(\mathcal{A}_{0,3}) = \operatorname{span}(1, e_\Omega) \cong \mathbf{C}$, and in the representation above, $e_\Omega = Q(iJ)U = iI$. Thus $\mathcal{A}_{0,3}$ can be considered as a four-dimensional complex algebra, where multiplication by i is multiplication by e_Ω.

Note that $j(\mathbf{R}_{0,3})$ is the three-dimensional real subspace of $M_2(\mathbf{C})$ consisting of all Hermitian matrices with zero trace:

$$j(\mathbf{R}_{0,3}) = \{T \in M_2(\mathbf{C}) : T = T^* \text{ and } \tau(T) = 0\}.$$

The elements $f_1 = e_3 e_2$, $f_2 = e_1 e_3$ and $f_3 = e_2 e_1$ generate $\mathcal{A}_{0,3}^+$, and $j(f_1) = \tau_1$, $j(f_2) = \tau_2$ and $j(f_3) = \tau_3$, where τ_1, τ_2 and τ_3 are the *associate Pauli matrices*. If $a = xI + v\tau_1 + u\tau_2 + y\tau_3 \in j(\mathcal{A}_{0,3}^+)$ then

$$a = \begin{bmatrix} x + iy & u + iv \\ -u + iv & x - iy \end{bmatrix} = \begin{bmatrix} z & w \\ -\bar{w} & \bar{z} \end{bmatrix},$$

where $z = x + iy$ and $w = u + iv$.

$$\text{Thus } j(\mathbf{A}_{0,3}^+) = \left\{ \begin{bmatrix} z & w \\ -\bar{w} & \bar{z} \end{bmatrix} : z, w \in \mathbf{C} \right\} = H.$$

Exercises

1. Suppose that

$$(S, T) = \left(\begin{bmatrix} a & b \\ c & d \end{bmatrix}, \begin{bmatrix} \alpha & \beta \\ \gamma & \delta \end{bmatrix} \right) \in M_2(\mathbf{R}) \times M_2(\mathbf{R}) \cong \mathcal{A}_{1,2}.$$

Calculate $(S,T)'$, $(S,T)^*$ and $\overline{(S,T)}$.

2. Let $i : \mathbf{R}_{1,1} \to \mathbf{R}_{1,2}$ be defined by $i((x,y)) = (x,y,0)$. Determine $j(\tilde{\imath}(\mathcal{A}_{1,1}))$.

3. Let $i : \mathbf{R}_{0,2} \to \mathbf{R}_{1,2}$ be defined by $i((x,y)) = (0,x,y)$. Determine $j(\tilde{\imath}(\mathcal{A}_{1,1}))$.

4. Let $i : \mathbf{R}_{0,2} \to \mathbf{R}_{0,3}$ be defined by $i((x,y)) = (x,0,y)$. Show that $j(\tilde{\imath}(\mathcal{A}_{0,2})) = M_2(\mathbf{R})$.

5. Suppose that

$$T = \begin{bmatrix} a & b \\ c & d \end{bmatrix} \in M_2(\mathbf{C}) \cong \mathcal{A}_{2,1}.$$

Calculate T', T^*, $\bar{T}$ and $\bar{T}T$.

7.6 Clifford algebras $\mathcal{A}(E,q)$ with $\dim E = 4$

The algebras $\mathcal{A}_4$ and $\mathcal{A}_{0,4}$

By Proposition 6.5.1 the algebras $\mathcal{A}_4$ and $\mathcal{A}_{0,4}$ are isomorphic as graded algebras, and $\mathcal{A}_4 \cong M_2(\mathbf{H})$. Also $\mathcal{A}_4^+ \cong \mathcal{A}_{0,4}^+ \cong A_3 \cong \mathbf{H} \oplus \mathbf{H}$.

We construct three explicit representations of $\mathcal{A}_4$.

First, define $j_1 : \mathbf{R}_4 \to M_2(\mathbf{H})$ by setting

$$j_1(e_1) = \boldsymbol{i} \otimes Q = \begin{bmatrix} 0 & \boldsymbol{i} \\ \boldsymbol{i} & 0 \end{bmatrix}, \quad j_1(e_2) = \boldsymbol{j} \otimes Q = \begin{bmatrix} 0 & \boldsymbol{j} \\ \boldsymbol{j} & 0 \end{bmatrix},$$

$$j_1(e_3) = \boldsymbol{k} \otimes Q = \begin{bmatrix} 0 & \boldsymbol{k} \\ \boldsymbol{k} & 0 \end{bmatrix}, \quad j_1(e_4) = -I \otimes J = \begin{bmatrix} 0 & 1 \\ -1 & 0 \end{bmatrix},$$

and extending by linearity. Then j_1 is a Clifford mapping, which extends to an algebra isomorphism $j_1 : \mathcal{A}_4 \to M_2(\mathbf{H})$. $j_1(\mathcal{A}_4^+)$ is generated by

$$j_1(e_2e_3) = \begin{bmatrix} \boldsymbol{i} & 0 \\ 0 & \boldsymbol{i} \end{bmatrix}, \quad j_1(e_3e_1) = \begin{bmatrix} \boldsymbol{j} & 0 \\ 0 & \boldsymbol{j} \end{bmatrix},$$

$$j_1(e_1e_2) = \begin{bmatrix} \boldsymbol{k} & 0 \\ 0 & \boldsymbol{k} \end{bmatrix}, \quad j(e_\Omega) = \begin{bmatrix} 1 & 0 \\ 0 & -1 \end{bmatrix},$$

so that in this representation

$$j_1(A_4^+) = \begin{bmatrix} \mathbf{H} & 0 \\ 0 & \mathbf{H} \end{bmatrix}.$$

Since $e_\Omega = I \otimes U$, it follows from Proposition 6.5.1 that we obtain a

Clifford mapping π_1 from $\mathbf{R}_{0,4}$ to $M_2(\mathbf{H})$ for which

$$\pi_1(e_1) = \boldsymbol{i} \otimes J = \begin{bmatrix} 0 & -\boldsymbol{i} \\ \boldsymbol{i} & 0 \end{bmatrix}, \quad \pi_1(e_2) = \boldsymbol{j} \otimes J = \begin{bmatrix} 0 & -\boldsymbol{j} \\ \boldsymbol{j} & 0 \end{bmatrix},$$

$$\pi_1(e_3) = \boldsymbol{k} \otimes J = \begin{bmatrix} 0 & -\boldsymbol{k} \\ \boldsymbol{k} & 0 \end{bmatrix}, \quad \pi_1(e_4) = I \otimes Q = \begin{bmatrix} 0 & 1 \\ 1 & 0 \end{bmatrix}.$$

This then extends to an algebra isomorphism of $\mathcal{A}_{0,4}$ onto $M_2(\mathbf{H})$.

Secondly, we can combine the representation j_1 with the representation of $\mathbf{H}$ as a subalgebra of $M_2(\mathbf{C})$, to obtain a representation j_2 of $\mathcal{A}_4$ in $M_4(\mathbf{C})$, where

$$j_2(e_1) = \tau_1 \otimes Q = iQ \otimes Q, \quad j_2(e_2) = \tau_2 \otimes Q = -J \otimes Q,$$
$$j_2(e_3) = \tau_3 \otimes Q = iU \otimes Q, \quad j_2(e_4) = -I \otimes J.$$

We shall see in the next section that this representation extends to a representation of $\mathcal{A}_5$.

Now $e_\Omega = -I \otimes U$, so that using Proposition 6.5.1 we obtain a representation $\pi_2 : \mathbf{R}_{0,4} \to M_4(\mathbf{C})$ for which

$$\pi_2(e_1) = \tau_1 \otimes J = iQ \otimes J, \quad \pi_2(e_2) = \tau_2 \otimes J = -J \otimes Q,$$
$$\pi_2(e_3) = \tau_3 \otimes J = iU \otimes J, \quad \pi_2(e_4) = -I \otimes Q.$$

We obtain a third Clifford mapping $j_3 : \mathbf{R}_4 \to M_4(\mathbf{C})$ by exchanging U and Q, setting

$$j_3(e_1) = \tau_1 \otimes U = iQ \otimes U, \quad j_3(e_2) = \tau_2 \otimes U = -J \otimes U,$$
$$j_3(e_3) = \tau_3 \otimes U = iU \otimes U, \quad j_3(e_4) = -iI \otimes Q,$$

and extending by linearity. We shall relate this to a representation of $\mathcal{A}_5^+$ in the next section.

The algebra $\mathcal{A}_{3,1}$

The Clifford algebras $\mathcal{A}_{3,1}$ and $\mathcal{A}_{1,3}$ are important for relativistic physics: we shall illustrate this in Chapter 9.

The algebra $\mathcal{A}_{3,1}$ is isomorphic to $M_2(\mathbf{H})$, and $\mathcal{A}_{3,1}^+ \cong \mathcal{A}_{2,1} \cong M_2(\mathbf{C})$. Let us construct an explicit representation. We obtain a Clifford mapping $j : \mathbf{R}_{3,1} \to M_2(\mathbf{H})$ by setting

$$j(e_1) = I \otimes J, \quad j(e_2) = \boldsymbol{i} \otimes U, \quad j(e_3) = \boldsymbol{j} \otimes U, \quad j(e_4) = I \otimes Q,$$

and extending by linearity. It then follows that

$$j(\mathcal{A}_{3,1}^+) = \left\{ \begin{bmatrix} a_1 + a_2\boldsymbol{k} & b_1\boldsymbol{i} + b_2\boldsymbol{j} \\ c_1\boldsymbol{i} + c_2\boldsymbol{j} & d_1 + d_2\boldsymbol{k} \end{bmatrix} : a_i, b_i, c_i, d_i \in \mathbf{R} \text{ for } i = 1, 2 \right\}.$$

This clearly does not reflect the fact that $\mathcal{A}_{3,1}^+$ is isomorphic to $M_2(\mathbf{C})$.

Since $\mathcal{A}_{3,2} \cong M_4(\mathbf{C})$, it is natural to represent $\mathcal{A}_{3,1}$ as a subalgebra of $M_4(\mathbf{C})$; for this we use the *Dirac matrices*. These are not only of practical use, but also of historical interest, since they feature in Dirac's work on the Dirac equation, which we shall discuss in Section 9.4. They are defined in terms of the Pauli spin matrices. Let

$$\gamma_0 = \sigma_0 \otimes Q = I \otimes Q = \begin{bmatrix} 0 & I \\ I & 0 \end{bmatrix},$$

$$\gamma_1 = \sigma_1 \otimes J = Q \otimes J = \begin{bmatrix} 0 & -Q \\ Q & 0 \end{bmatrix},$$

$$\gamma_2 = \sigma_2 \otimes J = iJ \otimes J = \begin{bmatrix} 0 & -iJ \\ iJ & 0 \end{bmatrix},$$

$$\gamma_3 = \sigma_3 \otimes J = U \otimes J = \begin{bmatrix} 0 & -U \\ U & 0 \end{bmatrix}.$$

In order to conform to our sign convention, we write γ_4 for γ_0. The mapping γ defined by

$$\gamma(x_1 e_1 + \cdots + x_4 e_4) = x_1 \gamma_1 + \cdots + x_4 \gamma_4$$

is then a Clifford mapping of $\mathbf{R}_{3,1}$ into $M_4(\mathbf{C})$, which extends to an isomorphism γ of $\mathcal{A}_{3,1}$ into $M_4(\mathbf{C})$. The image $\Gamma_{3,1} = \gamma(\mathcal{A}_{3,1})$ is called the *Dirac algebra*. Note that

$$\gamma(\mathbf{R}_{3,1}) = \left\{ \begin{bmatrix} \begin{bmatrix} 0 & 0 \\ 0 & 0 \end{bmatrix} & \begin{bmatrix} a & b \\ \bar{b} & d \end{bmatrix} \\ \begin{bmatrix} d & -b \\ -\bar{b} & a \end{bmatrix} & \begin{bmatrix} 0 & 0 \\ 0 & 0 \end{bmatrix} \end{bmatrix} : a, d \in \mathbf{R}, b \in \mathbf{C} \right\}$$

$$= \left\{ \begin{bmatrix} 0 & S \\ \mathrm{adj}S & 0 \end{bmatrix} : S = S^* \right\},$$

where $\mathrm{adj}S$ is the adjugate of S.

Let us now consider $\gamma(\mathcal{A}_{3,1}^+)$. Now

$$\gamma(e_1 e_3) = -J \otimes I, \quad \gamma(e_2 e_4) = -iJ \otimes U,$$
$$\gamma(e_1 e_4) = -Q \otimes U, \quad \gamma(e_2 e_3) = -iQ \otimes I,$$
$$\gamma(e_3 e_4) = -U \otimes U, \quad \gamma(e_1 e_2) = -iU \otimes I,$$

and $\gamma(e_\Omega) = -iI \otimes U$.

Suppose that

$$x = \lambda_0 I + \sum_{1 \le i < j \le 4} \lambda_{ij} e_i e_j + \lambda_\Omega e_\Omega \in \mathcal{A}_{3,1}^+.$$

Set

$$p = \lambda_0 + i\lambda_\Omega, \quad q = -(\lambda_{13} + i\lambda_{34}),$$
$$r = -(\lambda_{14} + i\lambda_{23}), \quad s = -(\lambda_{34} + i\lambda_{12}).$$

Then

$$\gamma(x) = \begin{bmatrix} pI & 0 \\ 0 & \bar{p}I \end{bmatrix} + \begin{bmatrix} qJ & 0 \\ 0 & \bar{q}J \end{bmatrix} + \begin{bmatrix} rQ & 0 \\ 0 & -\bar{r}Q \end{bmatrix} + \begin{bmatrix} sU & 0 \\ 0 & -\bar{s}U \end{bmatrix}$$

$$= \begin{bmatrix} \begin{bmatrix} a & b \\ c & d \end{bmatrix} & \begin{bmatrix} 0 & 0 \\ 0 & 0 \end{bmatrix} \\ \begin{bmatrix} 0 & 0 \\ 0 & 0 \end{bmatrix} & \begin{bmatrix} \bar{d} & -\bar{c} \\ -\bar{b} & \bar{a} \end{bmatrix} \end{bmatrix},$$

where

$$a = p + s, \quad b = -q + r, \quad c = q + r, \quad d = p - s.$$

The mapping $\tilde{\gamma} : \mathcal{A}_{3,1}^+ \to M_2(\mathbf{C})$ defined by

$$\tilde{\gamma}(x) = \begin{bmatrix} a & b \\ c & d \end{bmatrix}$$

is then an isomorphism of $\mathcal{A}_{3,1}^+$ onto $M_2(\mathbf{C})$.

The algebra $\mathcal{A}_{2,2}$

The Clifford algebra $\mathcal{A}_{2,2}$ is isomorphic to $M_4(\mathbf{R})$. We obtain a Clifford mapping $\mathbf{R}_{2,2} \to M_4(\mathbf{R})$ by setting

$$j(e_1) = I \otimes J, \quad j(e_2) = J \otimes Q, \quad j(e_3) = Q \otimes Q \text{ and } j(e_4) = U \otimes Q,$$

and extending by linearity. This then extends to an isomorphism of $\mathcal{A}_{2,2}$ onto $M_2(\mathbf{R})$. In detail, if $x = x_1 e_1 + x_2 e_2 + x_3 e_3 + x_4 e_4$ then

$$j(x) = \begin{bmatrix} 0 & u(x) \\ v(x) & 0 \end{bmatrix},$$

where

$$u(x) = \begin{bmatrix} -x_1 + x_4 & -x_2 + x_3 \\ x_2 + x_3 & -x_1 - x_4 \end{bmatrix}, \quad v(x) = \begin{bmatrix} x_1 + x_4 & -x_2 + x_3 \\ x_2 + x_3 & x_1 - x_4 \end{bmatrix}.$$

Further,

$$j(e_1e_2) = -J \otimes U, \quad j(e_1e_3) = -Q \otimes U, \quad j(e_1e_4) = -U \otimes U$$
$$j(e_3e_4) = J \otimes I, \quad j(e_2e_4) = Q \otimes I, \quad j(e_2e_3) = -U \otimes I,$$

and $j(e_\Omega) = I \otimes U$, so that

$$j(\mathcal{A}_{2,2}^+) = \left\{ \begin{bmatrix} x & 0 \\ 0 & y \end{bmatrix} : x, y \in M_2(\mathbf{R}) \right\}.$$

Thus if we set

$$\theta \left(\begin{bmatrix} x & 0 \\ 0 & y \end{bmatrix} \right) = (x, y) \in M_2(\mathbf{R}) \times M_2(\mathbf{R}),$$

then $\phi = \theta \circ j$ is an isomorphism of $\mathcal{A}_{2,2}^+$ onto $M_2(\mathbf{R}) \oplus M_2(\mathbf{R})$.

The algebra $\mathcal{A}_{1,3}$

The Clifford algebra $\mathcal{A}_{1,3}$ is isomorphic to $M_4(\mathbf{R})$, and $\mathcal{A}_{1,3}^+ \cong M_2(\mathbf{C})$. We can define a Clifford map of $\mathbf{R}_{1,3}$ into $M_4(\mathbf{R})$ by setting

$$j(e_1) = J \otimes Q, \quad j(e_2) = Q \otimes Q, \quad j(e_3) = U \otimes Q, \quad j(e_4) = I \otimes U,$$

and extending by linearity. Then $j(\mathcal{A}_{1,3}^+)$ is generated by

$$j(e_1e_4) = J \otimes J, \quad j(e_2e_4) = Q \otimes J \text{ and } j(e_3e_4) = U \otimes J.$$

Since $j(e_\Omega) = -I \otimes J$ is in the centre of $j(\mathcal{A}_{1,3}^+)$, we can identify $I \otimes J$ with i, and obtain the representation π of $\mathcal{A}_{1,3}^+$ onto $M_2(\mathbf{C})$ for which

$$\pi(e_1e_4) = iJ, \quad \pi(e_2e_4) = iQ, \quad \pi(e_3e_4) = iU,$$
$$\pi(e_2e_3) = J, \quad \pi(e_1e_3) = Q, \quad \pi(e_1e_2) = -U.$$

We have represented $\mathcal{A}_{3,1}$ as a subalgebra of $M_4(\mathbf{C})$; we can do the same for $\mathcal{A}_{1,3}$, using the associate Pauli matrices in place of the Pauli spin matrices. Let

$$\delta_0 = \tau_0 \otimes Q = I \otimes Q = \begin{bmatrix} 0 & I \\ I & 0 \end{bmatrix},$$

$$\delta_1 = \tau_1 \otimes J = iQ \otimes J = \begin{bmatrix} 0 & -iQ \\ iQ & 0 \end{bmatrix},$$

$$\delta_2 = \tau_2 \otimes J = -J \otimes J = \begin{bmatrix} 0 & J \\ -J & 0 \end{bmatrix},$$

$$\delta_3 = \tau_3 \otimes J = iU \otimes J = \begin{bmatrix} 0 & -iU \\ iU & 0 \end{bmatrix}.$$

To conform to this notation, we label the basis of $\mathbf{R}_{1,3}$ as e_0, e_1, e_2, e_3. The mapping δ defined by

$$\delta(x_0 e_0 + \cdots + x_3 e_4) = x_1 \delta_0 + \cdots + x_4 \delta_3$$

is then a Clifford mapping of $\mathbf{R}_{1,3}$ into $M_4(\mathbf{C})$, which extends to an isomorphism δ of $\mathcal{A}_{1,3}$ into $M_4(\mathbf{C})$.

Exercises

1. Let $j_1 : \mathcal{A}_4 \to M_2(\mathbf{H})$ be the representation defined above. Suppose that

$$a = \begin{bmatrix} h_1 & 0 \\ 0 & h_2 \end{bmatrix} \in j_1(\mathcal{A}_4^+).$$

   Show that $\bar{a} = \begin{bmatrix} \bar{h}_1 & 0 \\ 0 & \bar{h}_2 \end{bmatrix}$, and that $\bar{a}a = \begin{bmatrix} |h_1|^2 I & 0 \\ 0 & |h_2|^2 I \end{bmatrix}$.

2. Let $j_2 : \mathcal{A}_4 \to M_4(\mathbf{C})$ be the representation defined above. Show that

$$j_2(\mathcal{A}_4^+) = \begin{bmatrix} H & 0 \\ 0 & H \end{bmatrix},$$

   where H is the algebra used to define the quaternions.

3. Let $\gamma : \mathcal{A}_{3,1} \to M_4(\mathbf{C})$ be the representation defined above. Show that if

$$\gamma(x) = \begin{bmatrix} \begin{bmatrix} a & b \\ c & d \end{bmatrix} & \begin{bmatrix} 0 & 0 \\ 0 & 0 \end{bmatrix} \\ \begin{bmatrix} 0 & 0 \\ 0 & 0 \end{bmatrix} & \begin{bmatrix} \bar{d} & -\bar{c} \\ -\bar{b} & \bar{a} \end{bmatrix} \end{bmatrix},$$

then

$$\gamma(\bar{x}) = \begin{bmatrix} \begin{bmatrix} d & -b \\ -c & a \end{bmatrix} & \begin{bmatrix} 0 & 0 \\ 0 & 0 \end{bmatrix} \\ \begin{bmatrix} 0 & 0 \\ 0 & 0 \end{bmatrix} & \begin{bmatrix} \bar{a} & \bar{c} \\ \bar{b} & \bar{d} \end{bmatrix} \end{bmatrix}.$$

Show that

$$\gamma(\Delta(x)) = \begin{bmatrix} \theta I & 0 \\ 0 & \bar{\theta}I \end{bmatrix}, \quad \text{where } \theta = \det \begin{bmatrix} a & b \\ c & d \end{bmatrix}.$$

4. Let $\phi : \mathcal{A}_{2,2}^+ \to M_2(\mathbf{R}) \oplus M_2(\mathbf{R})$ be the isomorphism defined above. Suppose that

$$a = \lambda_0 I + \sum_{1 \le i < j \le 4} \lambda_{ij} e_i e_j + \lambda_\Omega e_\Omega \in \mathcal{A}_{2,2}^+.$$

Let

$$s_0 = \lambda_0 + \lambda_\Omega, \quad s_{12} = \lambda_{12} + \lambda_{34},$$
$$s_{13} = \lambda_{13} + \lambda_{24}, \quad s_{14} = \lambda_{14} + \lambda_{23},$$
$$d_0 = \lambda_0 - \lambda_\Omega, \quad d_{12} = \lambda_{12} - \lambda_{34},$$
$$d_{13} = \lambda_{13} - \lambda_{24}, \quad d_{14} = \lambda_{14} - \lambda_{23}.$$

Show that

$$\phi(a) = \left(\begin{bmatrix} s_0 - s_{14} & d_{12} - d_{13} \\ -d_{12} - d_{13} & s_0 + s_{14} \end{bmatrix}, \begin{bmatrix} d_0 + d_{14} & s_{12} + s_{13} \\ -d_{12} - d_{13} & d_0 - d_{14} \end{bmatrix} \right),$$

and

$$\phi(\bar{a}) = \left(\begin{bmatrix} s_0 + s_{14} & -d_{12} + d_{13} \\ d_{12} + d_{13} & s_0 - s_{14} \end{bmatrix}, \begin{bmatrix} d_0 - d_{14} & -s_{12} - s_{13} \\ d_{12} + d_{13} & d_0 + d_{14} \end{bmatrix} \right).$$

Let $\phi(a) = (S, T)$. Show that

$$\phi(\Delta(a)) = ((\det S)I, (\det T)I).$$

5. Let $\delta : \mathcal{A}_{1,3} \to M_4(\mathbf{C})$ be the representation of $\mathcal{A}_{1,3}$ defined above. Calculate $\delta(e_i e_j)$ for $0 \le i < j \le 3$ and $\delta(e_\Omega)$. Show that

$$\delta(\mathcal{A}_{1,3}^+) = \left\{ \begin{bmatrix} \begin{bmatrix} a & b \\ -\bar{b} & \bar{a} \end{bmatrix} & \begin{bmatrix} 0 & 0 \\ 0 & 0 \end{bmatrix} \\ \begin{bmatrix} 0 & 0 \\ 0 & 0 \end{bmatrix} & \begin{bmatrix} c & d \\ -\bar{d} & \bar{c} \end{bmatrix} \end{bmatrix} : a, b, c, d \in \mathbf{C} \right\}.$$

Compare this with $\gamma(\mathcal{A}_{3,1}^+)$.

7.7 Clifford algebras $\mathcal{A}(E,q)$ with $\dim E = 5$

Each of the Clifford algebras $\mathcal{A}_5$, $\mathcal{A}_{3,2}$ and $\mathcal{A}_{1,4}$ is isomorphic to $M_4(\mathbf{C})$. There are very many straightforward representations of these algebras, obtained by considering five suitable matrices of the form $\lambda S \otimes T$, where $\lambda = \pm 1$ or $\pm i$ and S and T are chosen from $\{I, Q, U, J\}$.

Each of the remaining algebras $\mathcal{A}_{4,1}$, $\mathcal{A}_{2,3}$ and $\mathcal{A}_{0,5}$ is not simple, and we shall also describe representations of the algebras $\mathcal{B}_{4,1}$, $\mathcal{B}_{2,3}$ and $\mathcal{B}_{0,5}$.

We shall only describe a few representations. We shall not give many details; we leave the reader to explore further. One worthwhile feature of the five-dimensional case is that the representations are extensions of interesting representations of Clifford algebras of four-dimensional spaces.

The algebra $\mathcal{A}_5$

We can define a Clifford mapping $j_1 : \mathbf{R}_5 \to M_4(\mathbf{C})$ by setting

$$
\begin{aligned}
j_1(e_1) &= \tau_1 \otimes Q = iQ \otimes Q, \quad j_1(e_2) = \tau_2 \otimes Q = -J \otimes Q, \\
j_1(e_3) &= \tau_3 \otimes Q = iU \otimes Q, \quad j_1(e_4) = -I \otimes J, \\
j_1(e_5) &= iI \otimes U,
\end{aligned}
$$

and extending by linearity. The corresponding representation j_1 extends the representation j_2 of $\mathcal{A}_4$ defined in the previous section.

The image of the even subalgebra $\mathcal{A}_5^+$ is generated by

$$
\begin{aligned}
j_1(e_4)j_1(e_1) &= \tau_1 \otimes U = iQ \otimes U, \quad j_1(e_4)j_1(e_2) = \tau_2 \otimes U = -J \otimes U, \\
j_1(e_4)j_1(e_3) &= \tau_3 \otimes U = iU \otimes U, \quad j_1(e_4)j_1(e_5) = -iI \otimes Q.
\end{aligned}
$$

Thus $j_1(\mathcal{A}_5^+) = j_3(\mathcal{A}_4)$, where j_3 is the representation of $\mathcal{A}_4$ defined in the previous section.

Another Clifford mapping $j_2 : \mathcal{A}_5 \to M_4(\mathbf{C})$ is obtained by setting

$$
\begin{aligned}
j_2(e_1) &= J \otimes I, \quad j_2(e_2) = iQ \otimes I, \\
j_2(e_3) &= iU \otimes U, \quad j_2(e_4) = U \otimes J, \\
j_2(e_5) &= iU \otimes Q,
\end{aligned}
$$

and extending by linearity. This then extends to an algebra isomorphism $j_2 : \mathcal{A}_5 \to M_4(\mathbf{C})$, which we shall use in Section 8.9.

The algebras $\mathcal{B}_{4,1}$ and $\mathcal{A}_{4,1}$

We define a Clifford mapping $k : \mathbf{R}_{4,1} \to M_2(\mathbf{H})$ by setting

$$k(e_1) = I \otimes J, \quad k(e_2) = \boldsymbol{i} \otimes U,$$
$$k(e_3) = \boldsymbol{j} \otimes U, \quad k(e_4) = \boldsymbol{k} \otimes U,$$
$$k(e_5) = I \otimes Q,$$

and extending by linearity, so that

$$k \left(\sum_{i=1}^{5} x_i e_i \right) = \left[\begin{array}{cc} x_2\boldsymbol{i} + x_3\boldsymbol{j} + x_4\boldsymbol{k} & -x_1 + x_5 \\ x_1 + x_5 & -x_2\boldsymbol{i} - x_3\boldsymbol{j} - x_4\boldsymbol{k} \end{array} \right].$$

Then k extends to an algebra homomorphism k of $\mathcal{A}_{4,1}$ onto $M_2(\mathbf{H})$. Since $k(e_\Omega) = -I$, the null-space of k is $\mathrm{span}(1 + e_\Omega)$, and $k(\mathcal{A}_{4,1}) \cong \mathcal{B}_{4,1}$.

We now define a Clifford mapping $j : \mathbf{R}_{4,1} \to M_2(\mathbf{H}) \oplus M_2(\mathbf{H})$ by setting $j(x) = (k(x), -k(x))$. This extends to an algebra isomorphism $j : \mathcal{A}_{4,1} \to M_2(\mathbf{H}) \oplus M_2(\mathbf{H})$. If $a \in \mathcal{A}_{4,1}$ then $j(a) = (k(a), k(a'))$.

If $E = \mathrm{span}(e_1, e_2, e_3, e_5)$, then E is isomorphic to $\mathbf{R}_{3,1}$, and the corresponding representation $j \circ \tilde{\imath} : \mathcal{A}(E) \to M_2(\mathbf{H})$ is the representation $j : \mathcal{A}_{3,1} \to M_2(\mathbf{H})$ given in the previous section.

The algebra $\mathcal{A}_{3,2}$

There are many interesting representations of $\mathcal{A}_{3,2}$. First let us extend the representation γ of $\mathcal{A}_{3,1}$ onto the Dirac algebra $\Gamma_{3,1}$. Let us set $\gamma_5 = I \otimes Q$. The mapping $\gamma : \mathbf{R}_{3,2} \to M_4(\mathbf{C})$ defined by

$$\gamma(x_1 e_1 + \cdots + x_5 e_5) = x_1 \gamma_1 + \cdots + x_5 \gamma_5$$

is then a Clifford mapping of $\mathbf{R}_{3,2}$ into $M_4(\mathbf{C})$, which extends to an algebra isomorphism γ of $\mathcal{A}_{3,2}$ onto $M_4(\mathbf{C})$. This isomorphism extends the representation γ of $\mathcal{A}_{3,1}$ onto $\Gamma_{3,1}$. Note that $\gamma(e_\Omega) = iI$, so that

$$M_4(\mathbf{C}) = \Gamma_{3,1} + e_\Omega \Gamma_{3,1} = \Gamma_{3,1} + i\Gamma_{3,1}.$$

Next, let us set

$$\delta_1 = iI \otimes Q, \ \delta_2 = iQ \otimes U, \ \delta_3 = iU \otimes U, \ \delta_4 = iJ \otimes U, \ \delta_5 = iI \otimes J.$$

The mapping $\delta : \mathbf{R}_{3,2} \to M_4(\mathbf{C})$ defined by

$$\delta(x_1 e_1 + \cdots + x_5 e_5) = x_1 \delta_1 + \cdots + x_5 \delta_5$$

is then a Clifford mapping of $\mathbf{R}_{3,2}$ into $M_4(\mathbf{C})$, which extends to an algebra isomorphism δ of $\mathcal{A}_{3,2}$ onto $M_4(\mathbf{C})$. Since $\delta(e_\Omega) = iI$ and $A_{3,2}^+$ is generated by $\{e_i e_\Omega : 1 \le i \le 5\}$, it follows that $\delta(A_{3,2}^+) = M_4(\mathbf{R})$.

Yet another representation can be obtained by setting

$$\epsilon_1 = I \otimes J, \ \epsilon_2 = iQ \otimes Q, \ \epsilon_3 = iU \otimes Q, \ \epsilon_4 = iJ \otimes Q, \ \epsilon_5 = iI \otimes U.$$

The mapping $\epsilon : \mathbf{R}_{3,2} \to M_4(\mathbf{C})$ defined by

$$\epsilon(x_1 e_1 + \cdots + x_5 e_5) = x_1 \epsilon_1 + \cdots + x_5 \epsilon_5$$

is then a Clifford mapping of $\mathbf{R}_{3,2}$ into $M_4(\mathbf{C})$, which extends to an algebra isomorphism ϵ of $\mathcal{A}_{3,2}$ onto $M_4(\mathbf{C})$.

The algebras $\mathcal{B}_{2,3}$ and $\mathcal{A}_{2,3}$

The non-simple Clifford algebra $\mathcal{A}_{2,3}$ is isomorphic to $M_4(\mathbf{R}) \oplus M_4(\mathbf{R})$. If we set

$$j_1(e_1) = I \otimes J, \quad j_1(e_2) = J \otimes Q,$$
$$j_1(e_3) = Q \otimes Q, \quad j_1(e_4) = U \otimes Q,$$
$$j_1(e_5) = I \otimes U,$$

and extend by linearity, we obtain a Clifford mapping of $\mathbf{R}_{2,3}$ into $M_4(\mathbf{R})$, which extends to an algebra homomorphism j_1 of $\mathcal{A}_{2,3}$ onto $M_4(\mathbf{R})$. Since $j(e_\Omega) = I$, the null-space of j_1 is $\mathrm{span}(I - e_\Omega)$, and j_1 factors as an isomorphism of $\mathcal{B}_{2,3}$ onto $M_4(\mathbf{R})$.

Note that $j_1 \otimes \tilde{\imath} : \mathcal{A}_{2,2} \to M_4(\mathbf{R})$ is the algebra isomorphism of $\mathcal{A}_{2,2}$ onto $M_4(\mathbf{R})$ defined in the previous section and that the mapping $j_1 \otimes \tilde{\imath} : \mathcal{A}_{1,3} \to M_4(\mathbf{R})$ is the algebra isomorphism of $\mathcal{A}_{1,3}$ onto $M_4(\mathbf{R})$ defined in the previous section.

Let us set $j_2(x) = (j_1(x), -j_1(x))$. Then j_2 is a Clifford mapping of $\mathbf{R}_{2,3}$ into $M_4(\mathbf{R}) \oplus M_4(\mathbf{R})$ which extends to an algebra homomorphism of $\mathcal{A}_{2,3}$ into $M_4(\mathbf{R}) \oplus M_4(\mathbf{R})$. Since $j_2(e_\Omega) = (I, -I)$, j_2 is injective. Thus j_2 is an isomorphism of $\mathcal{A}_{2,3}$ onto $M_4(\mathbf{R}) \oplus M_4(\mathbf{R})$. Further, $j_2(a) = (j_1(a), j_1(a'))$ for $a \in \mathcal{A}_{2,3}$.

The algebra $\mathcal{A}_{1,4}$

The algebra $\mathcal{A}_{1,4}$ is isomorphic to $M_4(\mathbf{C})$, and $\mathcal{A}_{1,4}^+$ is isomorphic to $M_2(\mathbf{H})$. There are various representations.

First we can start with the representation of $\mathcal{A}_5$, and use Proposition 6.5.1. Reversing the tensor products, changing some signs and rearranging, we obtain a representation j_1 for which

$$j_1(e_1) = iQ \otimes Q, \quad j_1(e_2) = iJ \otimes Q, \quad j_1(e_3) = U \otimes Q,$$
$$j_1(e_4) = iI \otimes J, \quad j_1(e_5) = I \otimes U.$$

Alternatively, we can extend the representation $j : \mathcal{A}_{1,3} \to M_4(\mathbf{R})$ of the previous section to obtain a representation j_2 for which

$$j_2(e_1) = J \otimes Q, \quad j_2(e_2) = Q \otimes Q, \quad j_2(e_3) = U \otimes Q,$$
$$j_2(e_4) = I \otimes U, \quad j_2(e_5) = iI \otimes J.$$

In the same way, we can extend the representation $\pi_2 : \mathcal{A}_{0,4} \to M_4(\mathbf{C})$ of the previous section to obtain a representation $j_3 : \mathcal{A}_{4,1} \to M_{(}\mathbf{C})$ for which

$$j_3(e_1) = iI \otimes U, \quad j_3(e_2) = \tau_1 \otimes J, \quad j_3(e_3) = \tau_2 \otimes J,$$
$$j_3(e_4) = \tau_3 \otimes J, \quad j_3(e_5) = I \otimes Q.$$

The algebras $\mathcal{A}_{0,5}$ and $\mathcal{B}_{0,5}$

The algebra $\mathcal{A}_{0,5}$ is not simple, and is isomorphic to $M_2(\mathbf{H}) \oplus M_2(\mathbf{H})$; the algebra $\mathcal{B}_{0,5}$ is isomorphic to $M_2(\mathbf{H})$. The easiest way to represent them is to extend the representations of $\mathcal{B}_{0,4}$ and $\mathcal{A}_{0,4}$.

We obtain a Clifford mapping k of $\mathbf{R}_{0,5}$ into $M_2(\mathbf{H})$ by setting

$$k(e_1) = i \otimes J, \quad k(e_2) = j \otimes J, \quad k(e_3) = k \otimes J,$$
$$k(e_4) = I \otimes Q, \quad k(e_5) = I \otimes U,$$

and extending by linearity. The mapping k then extends to an algebra homomorphism k of $\mathcal{A}_{0,5}$ onto $M_2(\mathbf{H})$. Since $j(e_\Omega) = -I$, the null space of this mapping is $1 + e_\Omega$, and k factors as an isomorphism of $\mathcal{B}_{0,5}$ onto $M_2(\mathbf{H})$. Define a Clifford mapping $j : \mathbf{R}_{0,5} \to M_2(\mathbf{H}) \oplus M_2(\mathbf{H})$ by setting $j(x) = (k(x), -k(x))$. This extends to an algebra homomorphism $j : \mathcal{A}_{0,5} \to M_2(\mathbf{H}) \oplus M_2(\mathbf{H})$. Since $j(e_\Omega) = (I, -I)$, j is an isomorphism of $\mathcal{A}_{0,5}$ onto $M_2(\mathbf{H}) \oplus M_2(\mathbf{H})$. Further, $j(a) = (k(a), k(a'))$ for $a \in \mathcal{A}_{0,5}$.

Exercises

1. Show that
$$\delta(\tilde{\imath}(\mathcal{A}_{3,1}^+)) = \left\{ \begin{bmatrix} S & -T \\ T & S \end{bmatrix} : S, T \in M_2(\mathbf{R}) \right\}.$$

2. Show that $\epsilon(\tilde{\imath}(\mathcal{A}_{3,1}^+))$ is generated by $iQ \otimes U$, $iJ \otimes U$ and $iU \otimes U$. Deduce that
$$\epsilon(\tilde{\imath}(\mathcal{A}_{3,1}^+)) = \left\{ \begin{bmatrix} S+iT & 0 \\ 0 & S-iT \end{bmatrix} : S, T \in M_2(\mathbf{R}) \right\} \cong M_2(\mathbf{C}).$$

3. Verify that the representation j_1 of $\mathcal{A}_{1,4}$ can be obtained from the representation of $\mathcal{A}_5$ in the way suggested.

7.8 The algebras $\mathcal{A}_6$, $\mathcal{B}_7$, $\mathcal{A}_7$ and $\mathcal{A}_8$

As the dimension becomes larger, constructions become more complicated. We end this chapter by considering representations of Clifford algebras for Euclidean spaces of dimensions 6, 7 and 8.

The algebra $\mathcal{A}_6$

We have constructed an isomorphism of $\mathcal{A}_5$ onto $M_4(\mathbf{C})$. Using the representation of $\mathbf{C}$ as a subalgebra of $M_2(\mathbf{R})$, we obtain a representation of $\mathcal{A}_5$ in $M_8(\mathbf{R})$, and we can extend this to a representation of $\mathcal{A}_6$. Thus, using the representation of $\mathcal{A}_5$ of the previous section, we obtain a Clifford mapping $j_6 : \mathbf{R}_6 \to M_8(\mathbf{R})$ by setting

$$
\begin{aligned}
j_6(e_1) &= J \otimes Q \otimes Q, \\
j_6(e_2) &= -I \otimes J \otimes Q, \\
j_6(e_3) &= J \otimes U \otimes Q, \\
j_6(e_4) &= -I \otimes I \otimes J, \\
j_6(e_5) &= J \otimes I \otimes U, \\
j_6(e_6) &= Q \otimes J \otimes U,
\end{aligned}
$$

and extending by linearity. This then extends to an algebra isomorphism $j_6 : \mathcal{A}_6 \to M_8(\mathbf{R})$. Note that $j_6(e_{\Omega_6}) = U \otimes J \otimes U$.

The algebra $\mathcal{B}_7$

In $\mathcal{A}_6$, $e_{\Omega_6}^2 = -1$ and $e_{\Omega_6} e_j = -e_j e_{\Omega_6}$ for $1 \le j \le 6$. We can therefore extend the Clifford mapping j_6 to a Clifford mapping $k_7 : \mathbf{R}_7 \to M_8(\mathbf{R})$ by setting $k_7(e_j) = j_6(e_j)$ for $1 \le j \le 6$ and $k_7(e_7) = j_6(e_{\Omega_6})$, and extending by linearity. This then extends to an algebra homomorphism $k_7 : \mathcal{A}_7 \to M_8(\mathbf{R})$. Then $k_7(e_{\Omega_7}) = -I$: k_7 is not injective, and its null-space is $\mathrm{span}(1 + e_{\Omega_7})$. The mapping k_7 therefore factors to give an algebra isomorphism of $\mathcal{B}_7$ onto $M_8(\mathbf{R})$.

The algebra $\mathcal{A}_7$

We can use the mapping k_7 to obtain a faithful representation of $\mathcal{A}_7$ in $M_8(\mathbf{R}) \oplus M_8(\mathbf{R})$. We define a Clifford mapping $j_7 : \mathbf{R}_7 \to M_{16}(\mathbf{R})$ by setting $j_7(x) = (k_7(x), -k_7(x))$ for $x \in \mathbf{R}_7$. This extends to an algebra homomorphism j_7 of $\mathcal{A}_7$ into $M_8(\mathbf{R}) \oplus M_8(\mathbf{R})$. Then $j_7(e_{\Omega_7}) = (I, -I)$, and j_7 is an isomorphism of $\mathcal{A}_7$ onto $M_8(\mathbf{R}) \oplus M_8(\mathbf{R})$. Further, if $a \in \mathcal{A}_7$

then

$$j_7(a) = (k_7(a), k_7(a')).$$

In a similar way, there is a representation j_7' of $\mathcal{A}_7$ in $M_{16}(\mathbf{R})$ for which

$$j_7'(x) = \left[\begin{array}{cc} 0 & k(x) \\ k(x) & 0 \end{array} \right], \quad \text{for } x \in \mathbf{R}_7.$$

The algebra $\mathcal{A}_8$

Finally we consider $\mathcal{A}_8$. We can obtain a representation of $\mathcal{A}_8$ by applying Proposition 6.5.1 to the representation of $\mathcal{A}_{4,4}$ given in Section 7.2. Instead, let us extend the representation j_7' of $\mathcal{A}_7$ described above. We set $j_8(e_j) = j_7'(e_j)$ for $1 \le j \le 7$, set $j_8(e_8) = I \otimes I \otimes I \otimes J$ and extend by linearity. Then j_8 is a Clifford mapping of $\mathbf{R}_8$ into $M_{16}(\mathbf{R})$ which extends to an algebra isomorphism $j_8 : \mathcal{A}_8 \to M_{16}(\mathbf{R})$. In this representation,

$$j_8(\mathcal{A}_8^+) = \left\{ \left[\begin{array}{cc} S & 0 \\ 0 & T \end{array} \right] : S, T \in M_8(\mathbf{R}) \right\} \cong M_8(\mathbf{R}) \oplus M_8(\mathbf{R}).$$

Exercise

1. Obtain a representation of $\mathcal{A}_8$ by applying Proposition 6.5.1 to the representation of $\mathcal{A}_{4,4}$ given in Section 7.2.

8

Spin

We have seen that a Clifford algebra $\mathcal{A}(E,q)$ is represented as an algebra of operators on a suitable spinor space. In particular, the group $\mathcal{G}(E,q)$ is represented as a group of linear isomorphisms of the spinor space. There is however a more important action, that of *adjoint conjugation*; we investigate this in the present chapter.

8.1 Clifford groups

Suppose that (E,q) is a regular quadratic space. We consider the action of $\mathcal{G}(E,q)$ on $\mathcal{A}(E,q)$ by *adjoint conjugation*. We set

$$Ad'_g(a) = gag'^{-1},$$

for $g \in \mathcal{G}(E,q)$ and $a \in \mathcal{A}(E,q)$.

We restrict attention to those elements of $\mathcal{G}(E,q)$ which stabilize E. The *Clifford group* $\Gamma = \Gamma(E,q)$ is defined as

$$\{g \in \mathcal{G}(E,q) : Ad'_g(x) \in E \text{ for } x \in E\}.$$

If $g \in \Gamma(E,q)$, we set $\alpha(g)(x) = Ad'_g(x)$. Then $\alpha(g) \in GL(E)$, and α is a homomorphism of Γ into $GL(E)$. α is called the *graded vector representation* of Γ.

In order to see why we do this, consider the case where y is an anisotropic element of E. Then $y' = -y$ and $yy' = q(y) \neq 0$, so that $y'^{-1} = y/q(y)$. Then

$$Ad'_y(y) = y^3/q(y) = -y,$$

while if $x \in y^{\perp}$ then

$$Ad'_y(x) = yxy/q(y) = -y^2x/q(y) = x.$$

Thus $y \in \Gamma$, and $\alpha(y)$ is the simple reflection in the direction y with mirror $y^{\perp}$.

Proposition 8.1.1 $\ker(\alpha) = \mathbf{R}^*$.

Proof   We use Proposition 5.4.1. Suppose that $g = g^+ + g^- \in \ker(\alpha)$. We show that $g^+ \in \mathbf{R}$ and that $g^- = 0$. If $x \in E$ then

$$g^+ x + g^- x = x g^+ - x g^-.$$

Since $g^+ x$ and $x g^+$ are in $\mathcal{A}^-$ and $g^- x$ and $x g^-$ are in $\mathcal{A}^+$, it follows that $g^+ x = x g^+$ and $g^- x = -x g^-$. Thus $g^+ \in Z(\mathcal{A})$. If d is even, then $g^+ \in \mathbf{R}$; if d is odd, then $g^+ \in Z(\mathcal{A}) \cap \mathcal{A}^+ = \mathbf{R}$.

Also, $g^- \in C_{\mathcal{A}}(\mathcal{A}^+) \cap \mathcal{A}^- = \mathrm{span}(1, e_\Omega) \cap \mathcal{A}^-$. If d is odd, then $\mathrm{span}(1, e_\Omega) \cap \mathcal{A}^- = \mathrm{span}(e_\Omega)$. But $e_\Omega x = x e_\Omega$ for all $x \in E$, and so $g^- = 0$. If d is even, then $\mathrm{span}(1, e_\Omega) \subseteq \mathcal{A}^+$, so that $g^- = 0$. $\square$

Proposition 8.1.2   *If $g \in \Gamma$ then $\bar{g} \in \Gamma$ and $\Delta(g) = \bar{g}g \in \ker(\alpha)$.*

Proof   The element $\bar{g}$ is invertible in $\mathcal{A}$; reversing the equation $gE = Eg'$ it follows that $\bar{g} \in \Gamma$. Suppose that $x \in E$ and $y = \alpha(g)(x)$. Reversing the equation $gx = yg'$, we find that $\bar{g}y = xg^* = x\bar{g}'$, so that $\alpha(\bar{g})y = x$ and $\alpha(\bar{g}g)(x) = x$. Thus $\bar{g}g \in \ker(\alpha)$. $\square$

Recall that $\mathcal{N}_*(E, q) = \{a \in \mathcal{A}(E, q) : \Delta(a) \in \mathbf{R}^*\}$.

Corollary 8.1.1 $\Gamma(E, q) \subseteq \mathcal{N}_*(E, q)$.

Proposition 8.1.3 $\alpha(\Gamma)$ *is the orthogonal group* $O(E, q)$.

Proof   Suppose that $g \in \Gamma$, $x \in E$ and $y = \alpha(g)(x)$. Then

$$q(y) = \Delta(y) = \Delta(gxg'^{-1}) = \Delta(g)\Delta(x)\Delta(g'^{-1}) = \Delta(x) = q(x).$$

Thus $\alpha(\Gamma) \subseteq O(E, q)$. But we have seen that the reflections are in $\alpha(\Gamma)$, and they generate $O(E, q)$, by the Cartan-Dieudonné theorem. $\square$

We therefore have the following exact sequence:

$$1 \longrightarrow \mathbf{R}^* \overset{\subseteq}{\longrightarrow} \Gamma(E, q) \overset{\alpha}{\longrightarrow} O(E, q) \longrightarrow 1.$$

Corollary 8.1.2   *The Clifford group $\Gamma(E, q)$ is the subgroup of $\mathcal{G}(E, q)$ generated by the anisotropic elements of E.*

Proof If $g \in \Gamma(E,q)$ then $\alpha(g) \in O(E,q)$, and so $\alpha(g)$ is the product of simple reflections. Thus there exist anisotropic elements $y_1, \ldots, y_n$ such that $\alpha(g) = \alpha(y_1) \ldots \alpha(y_n)$. Then $g y_n^{-1} \ldots y_1^{-1} \in \ker(\alpha) = \mathbf{R}^*$, so that $g = \lambda y_1 \ldots y_n$, for some $\lambda \in \mathbf{R}^*$.

If y is an anisotropic element of E then $y^2 = -q(y)$ and $y'y = q(y)$, so that any element of $\mathbf{R}^*$ can be written as the product of two anisotropic vectors. $\qquad\square$

8.2 Pin and Spin groups

It is customary to scale the elements of $\Gamma(E,q)$; we set

$$\text{Pin}(E,q) = \{g \in \Gamma(E,q) : \Delta(g) = \pm 1\},$$
$$\Gamma_1(E,q) = \{g \in \Gamma(E,q) : \Delta(g) = 1\}.$$

If (E,q) is a Euclidean space, then $\text{Pin}(E,q) = \Gamma_1(E,q)$; otherwise, $\Gamma_1(E,q)$ is a subgroup of $\text{Pin}(E,q)$ of index 2.

We have a short exact sequence

$$1 \longrightarrow D_2 \overset{\subseteq}{\longrightarrow} \text{Pin}(E,q) \overset{\alpha}{\longrightarrow} O(E,q) \longrightarrow 1;$$

$\text{Pin}(E,q)$ is a double cover of $O(E,q)$.

Theorem 8.2.1 $\text{Pin}(E,q)$ *is the subgroup of* $\mathcal{G}(E,q)$ *generated by the unit vectors (vectors x for which $q(x) = \pm 1$) in E.*

Proof Just as for Corollary 8.1.2. $\qquad\square$

In fact there is more interest in the subgroup $\text{Spin}(E,q)$ of $\text{Pin}(E,q)$ consisting of products of an even number of unit vectors in E. Thus $\text{Spin}(E,q) = \text{Pin}(E,q) \cap \mathcal{A}^+(E,q)$ and

$$\text{Spin}(E,q) = \{g \in \mathcal{A}^+(E,q) : gE = Eg \text{ and } \Delta(g) = \pm 1\}.$$

If x, y are unit vectors in E then $\alpha(xy) = \alpha(x)\alpha(y) \in SO(E,q)$, so that $\alpha(\text{Spin}(E,q)) \subseteq SO(E,q)$. Conversely, every element of $SO(E,q)$ is the product of an even number of simple reflections, and so $SO(E,q) \subseteq \alpha(\text{Spin}(E,q))$. Thus $\alpha(\text{Spin}(E,q)) = SO(E,q)$, and we have a short exact sequence

$$1 \longrightarrow D_2 \overset{\subseteq}{\longrightarrow} \text{Spin}(E,q) \overset{\alpha}{\longrightarrow} SO(E,q) \longrightarrow 1;$$

$\text{Spin}(E,q)$ is a double cover of $SO(E,q)$. Note also that if $a \in \text{Spin}(E,q)$

and $x \in E$ then $\alpha(a)(x) = axa^{-1}$; conjugation and adjoint conjugation by elements of $\mathrm{Spin}(E, q)$ are the same.

If (E, q) is a Euclidean space then $\Delta(a) = 1$ for all $a \in \mathrm{Spin}(E, q)$, and the same is the case if q is negative definite. Otherwise, the mapping $a \to \Delta(a)$ is a homomorphism of $\mathrm{Spin}(E, q)$ onto D_2. In this case, we set

$$\mathrm{Spin}^+(E, q) = \{a \in \mathrm{Spin}(E, q) : \Delta(a) = 1\} = \mathrm{Spin}(E, q) \cap \Gamma_1(E, q),$$

and $\mathrm{Spin}^-(E, q) = \{a \in \mathrm{Spin}(E, q) : \Delta(a) = -1\}$. $\mathrm{Spin}^+(E, q)$ is a normal subgroup of $\mathrm{Spin}(E, q)$ of index 2. Since $D_2 \subseteq \mathrm{Spin}^+(E, q)$, $\alpha(\mathrm{Spin}^+(E, q))$ is a normal subgroup of $SO(E, q)$ of index 2.

The group $\mathrm{Spin}(E, q)$ is a subgroup of $\mathcal{N}_{\pm 1}^+ = \{a \in \mathcal{A}^+ : \Delta a = \pm 1\}$. In low dimensions we have equality.

Proposition 8.2.1 *If $d \leq 5$ then $\mathrm{Spin}(E, q) = \mathcal{N}_{\pm 1}^+$.*

Proof Let A_3 be the space of trivectors in $\mathcal{A}(E, q)$ and let A_5 be the space of quinvectors. Thus $A_5 = \{0\}$ if $d < 5$ and $A_5 = \mathrm{span}(e_\Omega)$ if $d = 5$. Suppose that $g \in \mathcal{N}_{\pm 1}^+$ and that $x \in E$. Then $y = \alpha(g)x = gxg^{-1} \in \mathcal{A}^-$ and so we can write $y = \lambda + \mu + \nu$, with $\lambda \in E$, $\mu \in A_3$ and $\nu \in A_5$. Now $g^{-1} = \epsilon \bar{g}$, ,where $\epsilon = \pm 1$, so that $y = \epsilon gx\bar{g}$ and $\bar{y} = \epsilon g\bar{x}\bar{g} = -y$. But $\bar{y} = -\lambda + \mu - \nu$, so that $\mu = 0$. This deals with the case where $d \leq 4$.

Now suppose that $d = 5$. We can write $y = \lambda + \rho e_\Omega$, with $\rho \in \mathbf{R}$. Now $e_\Omega \in Z(\mathcal{A}(E, q))$, so that $y^2 = (\lambda^2 + \rho^2 e_\Omega^2) + 2\lambda \rho e_\Omega$. But

$$y^2 = gx^2 g^{-1} = -q(x)gg^{-1} = -q(x) \in \mathbf{R},$$

so that either $\lambda = 0$ or $\rho = 0$; thus either $\alpha(g)(x) \in E$ or $\alpha(g)(x) \in A_5$. Suppose, if possible, that $\alpha(g)(x) \in A_5$ for some $x \neq 0$. Since $\dim E_5 = 1$, and $\alpha(g)$ is injective, it follows that if x and z are linearly independent, then $\alpha(g)(z) \notin A_5$, and so $\alpha(g)(z) \in E$. But x and $x - z$ are also linearly independent, and so $\alpha(g)(x - z) \in E$.

But then $\alpha(g)(x) = \alpha(g)(x - z) + \alpha(g)(z) \in E$, giving a contradiction. Thus $y \in E$ and $g \in \mathrm{Spin}(E, q)$. $\qquad\square$

We shall see that this result does not hold when $d = 6$.

Exercise

1. Suppose that $g \in \mathrm{Spin}_d$. Show that there exist orthogonal unit vectors $f_1, \ldots, f_{2k}$ in $\mathbf{R}_d$ and $t_1, \ldots, t_k$ in $[0, 2\pi)$ such that

$$g = e^{t_1 f_1 f_2} \cdot \ldots \cdot e^{t_k f_{2k-1} f_{2k}},$$

and that every element $\mathcal{A}_d$ of this form is in Spin_d.

8.3 Replacing q by $-q$

In this section, we show that $\mathrm{Spin}(E, q)$ and $\mathrm{Spin}(E, -q)$ are isomorphic, and that their graded vector representations are essentially the same. We denote $\mathrm{Spin}(\mathbf{R}_d)$ by Spin_d and $\mathrm{Spin}(\mathbf{R}_{p,m})$ by $\mathrm{Spin}_{p,m}$. We have seen in Theorem 6.4.1 that $\mathcal{A}^+_{p+1,m} \cong \mathcal{A}_{p,m}$ and $\mathcal{A}^+_{p,m+1} \cong \mathcal{A}_{m,p}$. Thus if $p > 0$ then

$$\mathcal{A}^+_{p,m} \cong \mathcal{A}_{p-1,m} \cong \mathcal{A}^+_{m,p}.$$

We now construct an explicit isomorphism from $\mathcal{A}^+_{p,m}$ onto $\mathcal{A}^+_{m,p}$, and show that, using this isomorphism, the actions of $\mathrm{Spin}_{p,m}$ and $\mathrm{Spin}_{m,p}$ are naturally related by an *intertwining mapping*.

Suppose that $p + m = d$ and that $p \geq 1$. Let us denote the quadratic form on $\mathbf{R}_{p,m}$ by q and the quadratic form on $\mathbf{R}_{m,p}$ by r. Let us denote the standard orthogonal basis on $\mathbf{R}_{p-1,m}$ by $(e_1, \ldots, e_{d-1})$, the standard orthogonal basis on $\mathbf{R}_{p,m}$ by $(f_1, \ldots, f_d)$, and the standard orthogonal basis on $\mathbf{R}_{m,p}$ by $(g_1, \ldots, g_d)$. We define a bijective linear mapping T from $\mathbf{R}_{p,m}$ to $\mathbf{R}_{m,p}$ (the intertwining mapping) by reversing the order of terms; we set

$$T(x_1 f_1 + \cdots + x_d f_d) = x_d g_1 + x_{d-1} g_2 + \cdots + x_1 g_d = \sum_{j=1}^{d} x_{d+1-j} g_j,$$

so that $T(f_j) = g_{d+1-j}$ for $2 \leq j \leq d$. Then $r(T(x)) = -q(x)$.

Let $b_j = f_1 f_{j+1}$ for $1 \leq j \leq d-1$, and define a mapping θ from $\mathbf{R}_{p-1,m}$ to $\mathcal{A}^+_{p,m}$ by setting $\theta(\sum_{j=1}^{d-1} x_j e_j) = \sum_{j=1}^{d-1} x_j b_j$. Then θ is a Clifford mapping, which extends to an algebra isomorphism, again denoted by θ, from $\mathcal{A}_{p-1,m}$ onto $\mathcal{A}^+_{p,m}$.

Similarly, let $c_j = g_{d-j} g_d$ for $1 \leq j \leq d-1$, and define a mapping $\phi : \mathbf{R}_{p-1,m} \to \mathcal{A}^+_{p,m}$ by setting $\theta(\sum_{j=1}^{d-1} x_j e_j) = \sum_{j=1}^{d-1} x_j c_j$. Then ϕ is a Clifford mapping, which extends to an algebra isomorphism, again denoted by ϕ, from $\mathcal{A}_{p-1,m}$ onto $\mathcal{A}^+_{m,p}$.

Consequently, the mapping $\psi = \phi \circ \theta^{-1}$ is an algebra isomorphism from $\mathcal{A}^+_{p,m}$ onto $\mathcal{A}^+_{m,p}$, and $\psi(b_j) = c_j$, for $1 \leq j \leq d-1$.

Suppose now that $1 \leq i \leq d-1$ and that $1 \leq j \leq d$. Let $\epsilon_j = -1$ if $j = 1$ or $i + 1$ and let $\epsilon_j = 1$ otherwise. Then

$$\alpha(b_i)(f_j) = f_1 f_{i+1} f_j f_{i+1} f_1 / q(f_j) q(f_1) = \epsilon_j f_j.$$

Similarly,

$$\alpha(c_i)(T(f_j)) = g_{d-i}g_d g_{d+1-j}g_d g_{d-i}/r(g_d)r(g_{d-i}) = \epsilon_j g_{d+1-j} = \epsilon_j T(f_j).$$

Thus $\alpha(\psi(b_i)(T(f_j)) = \alpha(b_i)(f_j)$. By linearity, $\alpha(\psi(b_i)(T(y)) = \alpha(b_i)(y)$, for $y \in \mathbf{R}_{p,m}$. Since the elements b_i generate $\mathrm{Spin}_{p,m}$,

$$\alpha(\psi(g))(T(y)) = \alpha(g)(y) \text{ for } g \in \mathrm{Spin}_{p,m} \text{ and } y \in \mathbf{R}_{p,m}.$$

8.4 The spin group for odd dimensions

Suppose first that $p - m = 3$ (mod 4). Then $\mathcal{A}_{p,m}$ is not simple, and there exists a non-universal Clifford algebra $\mathcal{B}_{p,m}$. Let k be the corresponding Clifford mapping: from $\mathbf{R}_{p,m}$ to $\mathcal{B}_{p,m}$. If $x \in \mathbf{R}_{p,m}$, let $j(x) = (k(x), -k(x))$; then j is a Clifford mapping of $\mathbf{R}_{p,m}$ into $\mathcal{B}_{p,m} \oplus \mathcal{B}_{p,m}$ which extends to an algebraic isomorphism j of $\mathcal{A}_{p,m}$ onto $\mathcal{B}_{p,m} \oplus \mathcal{B}_{p,m}$. Also, $j(\mathcal{A}_{p,m}^+) = \{(b,b) : b \in \mathcal{B}_{p,m}\}$; if $a \in \mathcal{A}_{p,m}^+$ and $j(a) = (b,b)$ then $j(\Delta(a)) = (b^*b, b^*b)$. If $\pi_1 : \mathcal{B}_{p,m} \oplus \mathcal{B}_{p,m} \to \mathcal{B}_{p,m}$ is the projection onto the first co-ordinate, then $\phi = \pi_1 \circ j$ is a projection of $\mathcal{A}_{p,m}$ onto $\mathcal{B}_{p,m}$ whose restriction to $\mathcal{A}_{p,m}^+$ is an algebra isomorphism of $\mathcal{A}_{p,m}^+$ onto $\mathcal{B}_{p,m}$. Thus the restriction of ϕ to $\mathrm{Spin}_{p,m}$ is a group isomorphism of $\mathrm{Spin}_{p,m}$ onto $\phi(\mathrm{Spin}_{p,m})$. Now e_Ω is the product of an odd number of unit vectors, and so if x is a unit vector in $\mathbf{R}_{p,m}$ then $xe_\Omega \in \mathrm{Spin}_{p,m}$. Since $\phi(x) = \pm\phi(xe_\Omega)$, it follows that

$$\phi(\mathrm{Spin}_{p,m}) = \phi(\mathrm{Pin}_{p,m}) = \{b \in \mathcal{B}_{p,m} : b \text{ is the product of unit vectors}\}.$$

Further, if $a \in \mathrm{Spin}_{p,m}$ and $x \in \mathbf{R}_{p,m}$ then

$$\phi(\alpha(a))(x) = \phi(a)k(x)(\phi(a))^{-1}.$$

Thus instead of considering the action of $\mathrm{Spin}_{p,m}$ on $j(\mathbf{R}_{p,m})$, we can study the action of the group $\phi(\mathrm{Spin}_{p,m})$ (denoted by $\mathrm{Spin}_{p,m}(\mathcal{B})$) on $k(\mathbf{R}_{p,m})$.

When $p - m = 1$ (mod 4), $\mathrm{Spin}_{p,m} \cong \mathrm{Spin}_{m,p}$, and we can study the action of $\mathrm{Spin}_{m,p}$ in place of the action of $\mathrm{Spin}_{p,m}$.

8.5 Spin groups, for $d = 2$

$\mathrm{Spin}_2 \cong \mathrm{Spin}_{0,2} \cong \mathbf{T}$

The even Clifford algebra $\mathcal{A}_2^+$ has dimension 2 and is equal to $\mathrm{span}(1, e_\Omega)$. If $a = \lambda 1 + \mu e_\Omega \in \mathcal{A}_2^+$, then $\Delta(a) = \lambda^2 + \mu^2$; by Proposition 8.2.1

$$\mathrm{Spin}_2 = \mathcal{N}_1^+ = \{e^{te_\Omega} = \cos t.1 + \sin t.e_\Omega : 0 \le t < 2\pi\},$$

and the mapping $e^{it} \to e^{te_\Omega}$ is an isomorphism of $\mathbf{T}$ onto Spin_2. If $x \in \mathbf{R}_2$ then, by Proposition 2.1.3, $e^{te_\Omega} x = x e^{-te_\Omega}$, so that

$$\alpha(e^{te_\Omega})(x) = e^{te_\Omega} x e^{-te_\Omega} = e^{2te_\Omega} x.$$

Thus $\alpha(e^{te_\Omega}) = R_{2t}$, the rotation of $\mathbf{R}_2$ by an angle $2t$. The group $SO(2)$ is also isomorphic to $\mathbf{T}$, and we have the diagram

$$
\begin{array}{ccccccccc}
1 & \longrightarrow & D_2 & \overset{\subseteq}{\longrightarrow} & \mathrm{Spin}_2 & \overset{\alpha}{\longrightarrow} & SO(2) & \longrightarrow & 1 \\
 & & \cong \big\downarrow & & & & \big\downarrow \cong & & \\
1 & \longrightarrow & D_2 & \overset{\subseteq}{\longrightarrow} & \mathbf{T} & \overset{d}{\longrightarrow} & \mathbf{T} & \longrightarrow & 1,
\end{array}
$$

where $d(e^{i\theta}) = e^{2i\theta}$. Thus the double cover of $SO(2)$ is obtained by 'doubling the angle'.

$\mathrm{Spin}_{1,1}^+ \cong (\mathbf{R}, +) \times D_2 \cong (\mathbf{R}^*, \times)$ **and** $\mathrm{Spin}_{1,1} \cong (\mathbf{R}^*, \times) \times D_2$

We look at this in two ways. Again, $\mathcal{A}_{1,1}^+$ has dimension 2, and is equal to $\mathrm{span}(1, e_\Omega)$, but now $Q(\lambda 1 + \mu e_\Omega) = \lambda^2 - \mu^2$. By Proposition 8.2.1, $\mathrm{Spin}_2 = \mathcal{N}_{\pm 1}^+$.

First, let $S = \{e^{te_\Omega} = \cosh t.1 + \sinh t.e_\Omega : t \in \mathbf{R}\}$. Then S is a subgroup of $\mathcal{N}_1^+ = \mathrm{Spin}_{1,1}^+$, and the mapping $t \to e^{te_\Omega}$ is an isomorphism of $(\mathbf{R}, +)$ onto S. If $x \in \mathbf{R}_{1,1}$ then $\alpha(e^{te_\Omega})(x) = e^{2te_\Omega} x$. Further, $\mathcal{N}_1^+ = S \cup (-S)$, and $\alpha(s) = \alpha(-s)$; we have a short exact sequence

$$
1 \longrightarrow D_2 \overset{\subseteq}{\longrightarrow} \mathrm{Spin}_{1,1}^+ \overset{\alpha}{\longrightarrow} SO_c(1,1) \longrightarrow 1,
$$

where $SO_c(1,1)$ is the connected component of the identity in $SO(1,1)$.

Now $\Delta(e_\Omega) = Q(e_\Omega) = -1$, so that $e_\Omega \in \mathrm{Spin}_{1,1}^-$; if $x \in \mathbf{R}_{1,1}$ then $\alpha(e_\Omega)(x) = e_\Omega x e_\Omega = -x$, so that $\alpha(e_\Omega) = -I$. If $y \in \mathrm{Spin}_{1,1}^-$ then $y e_\Omega \in \mathrm{Spin}_{1,1}^+$ and $y = (y e_\Omega) e_\Omega$, so that

$$\mathrm{Spin}_{1,1} = \mathrm{Spin}_{1,1}^+ \cup \mathrm{Spin}_{1,1}^- = \mathrm{Spin}_{1,1}^+ \cup \mathrm{Spin}_{1,1}^+ e_\Omega.$$

Further, $e_\Omega y = y e_\Omega$ for $y \in \mathrm{Spin}_{1,1}^+$, and so $\mathrm{Spin}_{1,1} \cong \mathrm{Spin}_{1,1}^+ \times D_2$. We therefore have a short exact sequence

$$
1 \longrightarrow D_2 \overset{\subseteq}{\longrightarrow} \mathrm{Spin}_{1,1} \overset{\alpha}{\longrightarrow} SO(1,1) \longrightarrow 1.
$$

We now look at $\text{Spin}_{1,1}$ in a slightly different way; this essentially comes by considering the hyperbolic basis of $\mathbf{R}_{1,1}$. $\mathcal{A}_{1,1} \cong M_2(\mathbf{R})$. We realize this by using a different Clifford mapping to that given in Section 6.2. Let

$$j'(x_1, x_2) = -x_1 J + x_2 Q = \begin{bmatrix} 0 & x_1 + x_2 \\ -x_1 + x_2 & 0 \end{bmatrix}.$$

Then j' is a Clifford mapping, which extends to an algebra isomorphism, again denoted by j', of $\mathcal{A}_{1,1}$ onto $M_2(\mathbf{R})$. In this representation,

$$j'(e_\Omega) = U = \begin{bmatrix} 1 & 0 \\ 0 & -1 \end{bmatrix}.$$

Thus if $x = \lambda.1 + \mu.e_\Omega \in A_{1,1}^+$ then

$$j'(x) = \begin{bmatrix} \lambda + \mu & 0 \\ 0 & \lambda - \mu \end{bmatrix} = \begin{bmatrix} a & 0 \\ 0 & d \end{bmatrix},$$

say, and

$$j'(\bar{x}) = \lambda.1 - \mu.e_\Omega = \begin{bmatrix} \lambda - \mu & 0 \\ 0 & \lambda + \mu \end{bmatrix} = \begin{bmatrix} d & 0 \\ 0 & a \end{bmatrix},$$

so that $\Delta(x) = \lambda^2 - \mu^2 = ad$. In particular,

$$j'(e^{se_\Omega}) = \begin{bmatrix} e^s & 0 \\ 0 & e^{-s} \end{bmatrix}, \text{ so that } j'(S) = \left\{ \begin{bmatrix} a & 0 \\ 0 & a^{-1} \end{bmatrix} : a > 0 \right\}.$$

Thus

$$j'(\text{Spin}_{1,1}^+) = \left\{ \begin{bmatrix} a & 0 \\ 0 & a^{-1} \end{bmatrix} : a \in \mathbf{R}^* \right\} \cong (\mathbf{R}^*, \times)$$

and $j'(\text{Spin}_{1,1}) = \left\{ \begin{bmatrix} a & 0 \\ 0 & d \end{bmatrix} : ad = \pm 1 \right\} \cong (\mathbf{R}^*, \times) \times D_2.$

8.6 Spin groups, for $d = 3$

$\text{Spin}_3 \cong \text{Spin}_{0,3} \cong \mathbf{H}_1 = \{ h \in \mathbf{H} : \Delta(h) = 1 \} \cong S^3 \cong SU_2.$

We have seen that $\mathcal{B}_3 \cong \mathbf{H}$ and that the corresponding Clifford mapping k maps $\mathbf{R}_3$ onto the space $Pu(\mathbf{H})$ of pure quaternions. If $k : \mathcal{B}_3 \to \mathbf{H}$ is the corresponding algebra isomorphism, then $\Delta(k(b)) = k(b^*b)$, and so $k(\text{Spin}_3(\mathcal{B})) = \{ h \in \mathbf{H} : \Delta(h) = 1 \} = \mathbf{H}_1$. If $b \in \text{Spin}_3(\mathcal{B})$ and $x \in \mathbf{R}_3$

then $bxb^{-1} = k(b)k(x)(k(b))^{-1}$, so that the action of Spin_3 on $\mathbf{R}_3$ is the same as the action of $\mathbf{H}_1$ on $Pu(\mathbf{H})$ described in Theorem 4.9.1.

For an alternative approach, let us consider $\mathcal{A}_{0,3}$. We have seen in Section 7.5 that

$$\mathcal{A}_{0,3}^+ \cong \left\{ \begin{bmatrix} z & w \\ -\bar{w} & \bar{z} \end{bmatrix} : z, w \in \mathbf{C} \right\},$$

and that $\Delta(a) = |z|^2 + |w|^2$. Thus there is an isomorphism j of $\text{Spin}_{0,3}$ onto

$$\left\{ \begin{bmatrix} z & w \\ -\bar{w} & \bar{z} \end{bmatrix} : |z|^2 + |w|^2 = 1 \right\} = SU_2.$$

Recall that

$$j(\mathbf{R}_{0,3}) = \{ T \in M_2(\mathbf{C}) : T = T^* \text{ and } \tau(T) = 0 \}.$$

If $U \in SU_2$ and $T \in j(\mathbf{R}_{0,3})$, then $\alpha(U)(T) = UTU^*$; since

$$(UTU^*)^* = UT^*U^* = UTU^* \text{ and } \tau(UTU^*) = \tau(U^*UT) = \tau(T) = 0,$$

we have direct confirmation that $j(\mathbf{R}_{0,3})$ is invariant under conjugation by $SU(2)$.

$\text{Spin}_{2,1}^+ \cong \text{Spin}_{1,2}^+ \cong SL_2(\mathbf{R})$

In Section 7.5 we saw that $\mathcal{A}_{2,1} \cong M_2(\mathbf{C})$, that $\mathcal{A}_{2,1}^+ \cong \mathcal{A}_{1,1} \cong M_2(\mathbf{R})$, and that, under this isomorphism, if $a \in \mathcal{A}_{2,1}^+$ then $\Delta(a) = \det(a)$. Thus $\text{Spin}_{2,1}^+ \cong \text{Spin}_{1,2}^+ \cong SL_2(\mathbf{R})$.

Let us consider $\text{Spin}_{1,2}^+(\mathcal{B})$. Let $k : \mathcal{A}_{1,2} \to M_2(\mathbf{R})$ be the homomorphism defined in Section 7.5. Note that

$$k(e_1e_2) = -U, \quad k(e_1e_3) = Q, \quad k(e_2e_3) = J,$$

so that if $x = a1 + be_1e_2 + ce_1e_3 + de_2e_3 \in \mathcal{A}_{1,2}^+$ then

$$k(x) = \begin{bmatrix} a - b & c - d \\ c + d & a + b \end{bmatrix},$$

and $\det k(x) = (a^2 + d^2) - (b^2 + c^2) = \Delta(x)$.

Thus $k(\text{Spin}_{1,2}^+(\mathcal{B})) = SL_2(\mathbf{R})$, and the action of $\text{Spin}_{1,2}^+(\mathcal{B})$ on $\mathbf{R}_3$ corresponds to the conjugation action of $SL_2(\mathbf{R})$ on the three-dimensional subspace $M_2^{(0)}(\mathbf{R})$ of $M_2(\mathbf{R})$.

Exercise

1. If

$$T = \begin{bmatrix} a & b \\ c & d \end{bmatrix} \in SL_2(\mathbf{R}) \text{ and } Te_3T^{-1} = y_1e_1 + y_2e_2 + y_3e_3,$$

calculate y_1, y_2 and y_3.

8.7 Spin groups, for $d = 4$

$\mathrm{Spin}_4 \cong \mathrm{Spin}_{0,4} \cong \mathbf{H}_1 \times \mathbf{H}_1$

We consider the representation $j_1 : \mathcal{A}_{0,4} \to M_2(\mathbf{H})$ defined in Section 7.6. From the results there,

$$j_1(\mathrm{Spin}_{0,4}) = \begin{bmatrix} \mathbf{H}_1 & 0 \\ 0 & \mathbf{H}_1 \end{bmatrix}.$$

If

$$j_1(y) = \begin{bmatrix} k_1 & 0 \\ 0 & k_2 \end{bmatrix} \in j_1(\mathrm{Spin}_4) \text{ and } j_1(x) = \begin{bmatrix} 0 & -\bar{l} \\ l & 0 \end{bmatrix} \in j(\mathbf{R}_4),$$

then

$$j_1(yxy^{-1}) = \begin{bmatrix} 0 & -k_1\bar{l}\bar{k}_2 \\ k_2l\bar{k}_1 & 0 \end{bmatrix}.$$

Thus we can now recognize that this action is the same as the one described in Theorem 4.9.2.

$\mathrm{Spin}_{3,1}^{+} \cong \mathrm{Spin}_{1,3}^{+} \cong SL_2(\mathbf{C})$

Let us consider the representation γ of $\mathcal{A}_{3,1}$ as a subalgebra of $M_4(\mathbf{C})$ defined in Section 7.6. In this representation, $\gamma(\mathcal{A}_{3,1}^{+})$ is isomorphic to $M_2(\mathbf{C})$, and if $x \in \mathcal{A}_{3,1}$ then $\Delta(x) = 1$ if and only if $\det \tilde{\gamma}(x) = 1$. Thus $\mathrm{Spin}_{3,1}^{+} \cong SL_2(\mathbf{C})$. Further, we can identify $\mathbf{R}_{3,1}$ with the four-dimensional real subspace $\{T \in M_2(\mathbf{C}) : T = T^*\}$ of $M_2(\mathbf{C})$ consisting of Hermitian matrices. Then $\alpha(x)(y) = \tilde{\gamma}(x)y(\tilde{\gamma}(x))^{-1}$. This representation goes back to Dirac.

$\mathrm{Spin}_{2,2}^{+} \cong SL_2(\mathbf{R}) \times SL_2(\mathbf{R})$

Let us continue with the notation of Section 7.6. It follows from the results there that if $a \in \mathcal{A}_{2,2}^{+}$ and $\phi(a) = (S,T)$ then $\Delta(a) = 1$ if and only if $\det S = \det T = 1$. Thus $\mathrm{Spin}_{2,2}^{+} \cong SL_2(\mathbf{R}) \times SL_2(\mathbf{R})$. Further, if $a \in \mathrm{Spin}_{2,2}^{+}$, with $\phi(a) = (S,T)$, and $y \in \mathbf{R}_{2,2}$ then

$$j(\alpha(a)y) = \begin{bmatrix} S & 0 \\ 0 & T \end{bmatrix} \cdot \begin{bmatrix} 0 & u(y) \\ v(y) & 0 \end{bmatrix} \cdot \begin{bmatrix} S^{-1} & 0 \\ 0 & T^{-1} \end{bmatrix}$$

$$= \begin{bmatrix} 0 & Su(y)T^{-1} \\ Tv(y)S^{-1} & 0 \end{bmatrix}.$$

The mapping u is a linear isomorphism of $\mathbf{R}_{2,2}$ onto $M_2(\mathbf{R})$, and so the action of $\mathrm{Spin}_{2,2}^+$ is given by the equation $u(\alpha(a)y) = Su(y)T^{-1}$.

8.8 The group Spin$_5$

As the dimension d becomes larger, the spin groups become more complicated. For $d = 5$ we shall only consider the Euclidean case; we shall describe Spin$_5$, without describing its action on $\mathbf{R}_5$.

The group Spin$_5$ is a subgroup of $\mathcal{A}_5^+$, which is isomorphic to $M_2(\mathbf{H})$. We shall describe Spin$_5$ as a subgroup of $M_2(\mathbf{H})$; to do this, we need to consider $M_2(\mathbf{H})$ more carefully. We define an anti-automorphism, *conjugation*:

$$\text{if } T = \begin{bmatrix} a & b \\ c & d \end{bmatrix} \in M_2(\mathbf{H}) \text{ then } \bar{T} = \begin{bmatrix} \bar{a} & \bar{c} \\ \bar{b} & \bar{d} \end{bmatrix}.$$

The *hyper-unitary group* or *symplectic group* $HU(2)$ is then

$$HU(2) = \{T \in M_2(\mathbf{H}) : \bar{T}T = I\}.$$

(The term 'symplectic' was introduced by Hermann Weyl for groups which preserve an $\mathbf{H}$-Hermitian form.)

We shall show that Spin$_5$ is isomorphic to $HU(2)$.

Let l_1 be the isomorphism of $\mathcal{A}_4$ onto $\mathcal{A}_5^+$ defined in Theorem 6.4.1, let $j_1 : \mathcal{A}_4 \to M_2(\mathbf{H})$ be the isomorphism defined in Section 7.6, and let $\phi = j_1 \otimes l_1^{-1}$. Thus

$$\phi(e_1 e_2) = iQ = \begin{bmatrix} 0 & i \\ i & 0 \end{bmatrix}, \quad \phi(e_1 e_3) = jQ = \begin{bmatrix} 0 & j \\ j & 0 \end{bmatrix},$$

$$\phi(e_1 e_4) = kQ = \begin{bmatrix} 0 & k \\ k & 0 \end{bmatrix}, \quad \phi(e_1 e_5) = J = \begin{bmatrix} 0 & -1 \\ I & 0 \end{bmatrix}.$$

Now $\overline{e_1 e_i} = -e_1 e_i$ and $\overline{\phi(e_1 e_i)} = -\phi(e_1 e_i)$ for $2 \le i \le 5$, so that $\overline{\phi(e_1 e_i)} = \overline{e_1 e_i}$.

If $2 \leq i < j \leq 5$ then

$$\overline{\phi(e_ie_j)} = \overline{\phi(e_1e_ie_1e_j)} = \overline{\phi(e_1e_i)\phi(e_1e_j)}$$
$$= \overline{\phi(e_1e_j)}.\overline{\phi(e_1e_i)} = \phi(\overline{e_1e_j}).\phi(\overline{e_1e_i}) = \phi(e_1e_j)\phi(e_1e_i)$$
$$= \phi(e_1e_je_1e_i) = \phi(-e_ie_j) = \phi(\overline{e_ie_j}).$$

As a result, if $1 \leq i < j < l < m \leq 5$ then

$$\overline{\phi(e_ie_je_le_m)} = \overline{\phi(e_ie_j)\phi(e_le_m)} = \overline{\phi(e_le_m)}.\overline{\phi(e_ie_j)}$$
$$= \phi(\overline{e_le_m}).\phi(\overline{e_ie_j}) = \phi(\overline{e_le_m}.\overline{e_ie_j}) = \phi(\overline{e_ie_je_le_m}).$$

Thus if $a \in \mathcal{A}_5^+$ then $\phi(\bar{a}) = \overline{\phi(a)}$. Thus $a \in \mathrm{Spin}_5 = \mathcal{N}_1^+$ if and only if $\phi(a) \in HU(2)$.

Exercise

1. If $h = (h_1, h_2)$ and $k = (k_1, k_2)$ are in $\mathbf{H}^2$, let $\langle h, k \rangle = \bar{h}_1k_1 + \bar{h}_2k_2$: the mapping $(h, k) \to \langle h, k \rangle$ is the *standard* $\mathbf{H}$*-Hermitian form* on $\mathbf{H}^2$. Show that $T \in HU(2)$ if and only if $\langle T(h), T(k) \rangle = \langle h, k \rangle$ for all $h, k \in \mathbf{H}^2$.

8.9 Examples of spin groups for $d \geq 6$

When $d \geq 5$ it is no longer the case that $\mathrm{Spin}_{p,m}^+ = \mathcal{N}_1^+(\mathbf{R}_{p,m})$, so that we need to work harder. We shall only consider Spin_6 in any detail.

Theorem 8.9.1 $\mathrm{Spin}_6 \cong SU_4$.

Proof Let l_1 be the isomorphism of $\mathcal{A}_5$ onto $\mathcal{A}_6^+$ defined in Theorem 6.4.1, let $j_2 : \mathcal{A}_5 \to M_4(\mathbf{C})$ be the isomorphism defined in Section 7.7, and let $\phi = j_2 \otimes l_1^{-1}$. Thus

$$\phi(e_1e_2) = J \otimes I, \quad \phi(e_1e_3) = iQ \otimes I,$$
$$\phi(e_1e_4) = iU \otimes U, \quad \phi(e_1e_5) = U \otimes J,$$
$$\phi(e_1e_6) = iU \otimes Q, \quad \phi(e_\Omega) = i.$$

We consider the usual conjugation on $M_4(\mathbf{C})$: $(T^*)_{ij} = \overline{T_{ji}}$. Now $\overline{e_1e_i} = (e_1e_i)^* = -e_1e_i$, and $(\phi(e_1e_i))^* = -\phi(e_1e_i)$, for $2 \leq i \leq 6$. Arguing as in the previous section, it follows that ϕ is a *-isomorphism of $\mathcal{A}_6^+$ onto $M_4(\mathbf{C})$. Thus

$$\phi(\mathcal{N}^+) = \{A \in M_4(\mathbf{C}) : A^*A = \lambda I \text{ with } \lambda \in \mathbf{R}^*\},$$

and $\phi(\mathcal{N}_1^+) = U_4$.

First we use a connectedness argument to show that $\mathrm{Spin}_6 \subseteq SU(4)$. If x and y are unit vectors in $\mathbf{R}_6$ then $(xy)^2 = -1$, so that $\det \phi(xy) = \pm 1$, and so $\det \phi(\mathrm{Spin}_6) \subseteq \{1, -1\}$. The set of unit vectors in $\mathbf{R}^6$ is connected, and so the set

$$P_2 = \{a \in \mathcal{A}_6 : a = xy \text{ with } x, y \text{ unit vectors in } \mathbf{R}^6\}$$

is connected in $\mathcal{A}_6$. Thus the function $xy \to \det \phi(xy)$ is constant on P_2. But if x is a unit vector then $\det \phi(x^2) = \det(-I) = 1$. Thus $\phi(P_2) \subseteq SU(4)$. Since Spin_6 is generated by P_2, it follows that $\phi(\mathrm{Spin}_6) \subseteq SU_4$.

Since $\phi(\mathrm{Spin}_6) \subseteq SU_4$ and $\phi(\mathcal{N}_1^+) = U_4$, they are certainly not equal. In order to go further, we need to understand the relation between $\mathcal{N}_1^+$ and Spin_6. We consider e^{te_Ω} for $t \in \mathbf{R}$. Since $e_\Omega^2 = -1$, the mapping $e^{it} \to e^{te_\Omega}$ is an isomorphism of $\mathbf{T}$ onto a subgroup G of $\mathcal{N}_1^+$. Note that $\phi(e^{te_\Omega}) = e^{it}I$, so that $\det \phi(e^{te_\Omega}) = e^{4it}$.

If $x \in \mathbf{R}_6$ then $e^{te_\Omega}xe^{-te_\Omega} = e^{2te_\Omega}x$, so that $e^{te_\Omega} \in \mathrm{Spin}_6$ if and only if $t = j\pi/2$ for some $j \in \mathbf{Z}$. Further,

$$e^{te_\Omega} = 1 \quad \text{if } t = 2k\pi,$$
$$e^{te_\Omega} = e_\Omega \quad \text{if } t = 2k\pi + \pi/2,$$
$$e^{te_\Omega} = -1 \quad \text{if } t = (2k+1)\pi,$$
$$e^{te_\Omega} = -e_\Omega \quad \text{if } t = 2k\pi + 3\pi/2.$$

Lemma 8.9.1 *If $g \in N_1$ and $x \in \mathbf{R}_6$ then $gxg^{-1} = e^{te_\Omega}u$ for some $e^{te_\Omega} \in G$ and $u \in \mathbf{R}_6$.*

Proof Arguing as in Proposition 8.2.1, we can write $gxg^{-1} = \lambda + e_\Omega\rho$, where $\lambda, \rho \in \mathbf{R}_6$. Then

$$(gxg^{-1})^2 = gx^2g^{-1} = (\lambda + e_\Omega\rho)^2 = \lambda^2 + \rho^2 + e_\Omega(\rho\lambda - \lambda\rho) \in \mathbf{R},$$

so that $\lambda\rho = \rho\lambda$. If $\lambda \neq 0$, we can write $\rho = \alpha\lambda + \sigma$, where $\sigma \perp \lambda$. Then

$$\lambda\sigma + \sigma\lambda = 0 \text{ and } \lambda\sigma - \sigma\lambda = \lambda\rho - \rho\lambda = 0,$$

so that $\lambda\sigma = 0$ and $\sigma = 0$. Thus λ and ρ are linearly dependent, and we can write $\lambda = \cos t.u$ and $\rho = \sin t.u$, for some $0 \leq \theta < 2\pi$ and $u \in \mathbf{R}^6$. Then $gxg^{-1} = e^{te_\Omega}u = e^{te_\Omega/2}ue^{-te_\Omega/2}$. $\qquad\square$

Lemma 8.9.2 $\mathcal{N}_1^+ = G.\mathrm{Spin}_6$.

Proof Suppose that $g \in \mathcal{N}_1^+$. Let x be a non-zero element of $\mathbf{R}_6$. By the preceding lemma, there exist $e^{te_\Omega} \in G$ and $u \in \mathbf{R}_6$ such that $gxg^{-1} =$

$e^{te_\Omega/2} u e^{-te_\Omega/2}$. Let $f = e^{-te_\Omega/2} g$; then $f x f^{-1} = u \in \mathbf{R}_6$. We shall show that $f \in \mathrm{Spin}_6$, so that $g = e^{te_\Omega/2} f \in G. \mathrm{Spin}_6$.

Suppose that y is linearly independent of x. By the preceding lemma,

$$ f y f^{-1} = e^{se_\Omega} v e^{-se_\Omega} = \cos 2s.v + \sin 2s.e_\Omega v, $$

for some $s \in \mathbf{R}$ and $v \in \mathbf{R}_6$. We shall show that $\sin 2s = 0$, so that $f y f^{-1} \in \mathbf{R}_6$, and $f \in \mathrm{Spin}_6$. Suppose, if possible, that $\sin 2s \neq 0$. Now

$$ f(x+y)f^{-1} = e^{re_\Omega} w e^{-re_\Omega} = \cos 2r.w + \sin 2r.e_\Omega w, $$

for some $r \in \mathbf{R}$ and $w \in \mathbf{R}_6$. Since $f(x+y)f^{-1} = u + f e_2 f^{-1}$ it follows that

$$ \cos 2r.w = u + cos2s.v \quad \text{and} \quad \sin 2r.w = \sin 2s.v. $$

Thus v and w are linearly dependent. Since $u = \cos 2r.w - \cos 2s.v$, u and v are also linearly dependent. Thus $f y f^{-1} = \alpha(\cos 2s.u + \sin 2s.e_\Omega u)$, for some $\alpha \in \mathbf{R}^*$. Let $y_1 = y - \alpha \cos 2s.x$. Then

$$ f y_1 f^{-1} = \alpha \sin 2s.e_\Omega u \in e_\Omega \mathbf{R}_6. $$

Suppose now that z is an element of $\mathbf{R}_6$ linearly independent of x and y. If $t = f z f^{-1} \notin \mathbf{R}_6$, then, using the same argument as before, $t \in \mathrm{span}(u, e_\Omega u) = f\mathrm{span}(x,y)f^{-1}$. This is not possible, and so $t \in \mathbf{R}_6$. But now we can repeat the argument, replacing x by z, to deduce that

$$ y_1 \in \mathrm{span}(t, e_\Omega t) \cap e_\Omega \mathbf{R}_6 = \mathrm{span}(e_\Omega t). $$

Thus t is a multiple of u, which is again impossible. We therefore conclude that $\sin 2s = 0$. ☐

We now complete the proof of the theorem. If $g \in \mathcal{N}_1^+$ and $g = e^{te_\Omega} f$, with $f \in \mathrm{Spin}_6$, then $\det \phi(g) = \det \phi(e^{te_\Omega}) \det \phi(f) = e^{4it}$, so that if $\phi(g) \in SU(4)$ then $e^{it} = i, -1, -i$ or 1, so that $e^{te_\Omega} = e_\Omega, -1, -e_\Omega$ or 1, and $g \in \mathrm{Spin}_6$. Thus $\phi(\mathrm{Spin}_6) = SU(4)$. ☐

This algebraic proof also gives information about the relationship between Spin_6 and $\mathcal{N}_1^+$. There is however a shorter argument, based on dimension, which gives a shorter proof of the theorem (but no more). As we have seen, $\phi(\mathrm{Spin}_6)$ is a subgroup of SU_4. Since Spin_6 is a double cover of SO_6, it has the same dimension as SO_6, which is $6.5/2 = 15$. But U_4 has dimension $4^2 = 16$, and so SU_4 has dimension 15 as well, so that $\phi(\mathrm{Spin}_6)$ is an open-and-closed subgroup of SU_4. But SU_4 is connected, and so $\phi(\mathrm{Spin}_6) = SU_4$.

Spin$_7$ is isomorphic to a subgroup of SO_8

We know that $\mathcal{A}_7^+ \cong \mathcal{A}_6$ and that $\mathcal{A}_6 \cong \mathcal{A}_{2,4} \cong M_4(\mathcal{A}_{0,2}) \cong M_8(\mathbf{R})$. Following through the usual calculations, we find that there is an isomorphism $k : \mathcal{A}_7^+ \to M_8(\mathbf{R})$ such that

$$k(e_1 e_7) = J \otimes I \otimes I, \quad k(e_2 e_7) = U \otimes J \otimes I, \quad k(e_3 e_7) = Q \otimes J \otimes Q,$$
$$k(e_4 e_7) = Q \otimes J \otimes U, \quad k(e_5 e_7) = Q \otimes I \otimes J, \quad k(e_6 e_7) = U \otimes Q \otimes J.$$

Since $k(\overline{e_i e_7}) = -k(e_i e_7) = (k(e_i e_7))'$ for $1 \leq i \leq 6$, it follows that $k(\bar{a}) = (k(a))'$ for $a \in \mathcal{A}_7^+$, so that $k(\mathcal{N}_1^+) = O(8)$. The connectedness argument that we used above then shows that $k(\text{Spin}_7) \subseteq SO(8)$.

Note that $\dim(\text{Spin}_7) = \dim(SO(7)) = 21$, while $\dim(SO(8)) = 28$. Thus several constraints need to be imposed on an element of $\mathcal{N}_1^+$ for it to belong to Spin$_7$.

8.10 Table of results

Let us tabulate the results that have been obtained.

Spin groups

Spin$_2 \cong$ Spin$_{0,2}$	**T**
Spin$_{1,1}^+$	$(\mathbf{R}, +) \times D_2 \cong (\mathbf{R}^*, \times)$
Spin$_{1,1}$	$(\mathbf{R}^*, \times) \times D_2$
Spin$_3 \cong$ Spin$_{0,3}$	$\mathbf{H}_1 \cong S^3 \cong SU_2$
Spin$_{2,1}^+ \cong$ Spin$_{1,2}^+$	$SL_2(\mathbf{R})$
Spin$_4 \cong$ Spin$_{0,4}$	$\mathbf{H}_1 \times \mathbf{H}_1$
Spin$_{2,2}^+$	$SL_2(\mathbf{R}) \times SL_2(\mathbf{R})$
Spin$_5 \cong$ Spin$_{0,5}$	$HU(2)$
Spin$_6 \cong$ Spin$_{0,6}$	SU_4
Spin$_7 \cong$ Spin$_{0,7}$	A proper subgroup of SO_8

PART THREE

SOME APPLICATIONS

9

Some applications to physics

In this chapter, we briefly describe some applications of Clifford algebras to physics. The first of these concerns the spin of an elementary particle, when the spin is $1/2$. Pauli introduced the Pauli spin matrices to describe this phenomenon. Physics is concerned with partial differential operators: building on Pauli's ideas, Dirac introduced a first order differential operator, now called the Dirac operator, in order to formulate the relativistic wave equation for an electron; this in turn led to the discovery of the positron. We shall study the Dirac operator more fully in the next chapter. Here we show how it can be used to formulate Maxwell's equation for an electromagnetic field in a particularly simple way, and will also describe the Dirac equation.

9.1 Particles with spin $1/2$

The Pauli spin matrices were introduced by Pauli to represent the internal angular momentum of particles which have spin $1/2$. Let us briefly describe how this can be interpreted in terms of the Clifford algebra $\mathcal{A}_{0,3}$. In quantum mechanics, an *observable* corresponds to a Hermitian linear operator T on a Hilbert space H, and, when the possible values of the observable are discrete, these possible values are the eigenvalues of T.

The Stern-Gerlach experiment showed that elementary particles have an intrinsic angular momentum, or *spin*. If x is a unit vector in $\mathbf{R}^3$ then the component J_x of the spin in the direction x is an observable. In a non-relativistic setting, this leads to the consideration of a linear mapping $x \to J_x$ from the Euclidean space $V = \mathbf{R}^3$ into the space $L_h(H)$ of Hermitian operators on an appropriate state space H. Particles are either *bosons*, in which case the eigenvalues of J_x are integers, or *fermions*, in which case the eigenvalues of J_x are of the form $(2k+1)/2$. In the case where the particle has spin $1/2$, each of the operators J_x has just two eigenvalues, namely $1/2$ (*spin up*) and $-1/2$ (*spin down*). Consequently, $J_x^2 = I$.

This equation suggests that we consider the negative-definite quadratic form $q(x) = -\|x\|^2$ on $\mathbf{R}^3$, and consider $\mathbf{R}_{0,3}$ as a subspace of $\mathcal{A}_{0,3}$. Now let $j : \mathcal{A}_{0,3} \to M_2(\mathbf{C})$ be the isomorphism defined in Section 7.5, and let $J_i = \frac{1}{2}j(e_i) = \frac{1}{2}\sigma_i$ for $i = 1, 2, 3$. Thus

$$ J_1 J_2 = (i/2)J_3, \quad J_2 J_3 = (i/2)J_1, \quad J_3 J_1 = (i/2)J_2. $$

Now the Pauli spin matrices are Hermitian, and identifying $\mathcal{A}_{0,3}$ with $M_2(\mathbf{C})$, we see that we can take the state space H to be the spinor space $\mathbf{C}^2$. If $x = (x_1, x_2, x_3) \in \mathbf{R}_{0,3}$ and $q(x) = -1$ then

$$ J_x = \frac{1}{2} \begin{bmatrix} x_3 & x_1 - ix_2 \\ x_1 + ix_2 & -x_3 \end{bmatrix} $$

is a Hermitian matrix with eigenvalues $1/2$ and $-1/2$; the corresponding eigenvectors are

$$ \begin{bmatrix} x_1 - ix_2 \\ 1 - x_3 \end{bmatrix} \text{ and } \begin{bmatrix} x_1 - ix_2 \\ -1 - x_3 \end{bmatrix} $$

respectively.

9.2 The Dirac operator

We now begin to study the analytic use of Clifford algebras. Some knowledge of analysis is required. In particular, familiarity with the differential calculus for vector-valued functions of several real variables is needed. If U is an open subset of a finite-dimensional real vector space and F is a finite-dimensional vector space then we define $C(U, F)$ to be the vector space of all continuous F-valued functions defined on U, and, for $k > 0$, we define $C^k(U, F)$ to be the vector space of all k-times continuously differentiable F-valued functions defined on U.

Suppose that U is an open subset of Euclidean space $\mathbf{R}_d$, that F is a finite-dimensional vector space and that $f \in C^2(U, F)$. Then the second-order linear differential operator $\Delta : C^2(U, F) \to C(U, F)$ (the

Laplacian) is defined as

$$\Delta f = \sum_{j=1}^{d} \frac{\partial^2 f}{\partial x_j^2}.$$

(Once again, conventions vary: for some authors, $\Delta f = -\sum_{j=1}^{d}(\partial^2 f/\partial x_j^2)$, and for others, a factor $\frac{1}{2}$ is included in the definition.) A *harmonic* function is one which satisfies $\Delta f = 0$. The study of the Laplacian, and of harmonic functions, is one of the major topics of analysis.

We can consider other quadratic forms on $\mathbf{R}^d$. Suppose that $\mathbf{R}_{p,m}$ is the standard regular quadratic space with signature (p, m). Then we define the corresponding Laplacian Δ_q as

$$\Delta_q = \sum_{j=1}^{d} q(e_j) \frac{\partial^2 f}{\partial x_j^2} = \sum_{j=1}^{p} \frac{\partial^2 f}{\partial x_j^2} - \sum_{j=p+1}^{p+m} \frac{\partial^2 f}{\partial x_j^2},$$

and define *q-harmonic functions* f as those for which $\Delta_q f = 0$.

It was Dirac who first saw that the second-order linear differential operator $-\Delta_q$ can be written as the square of a first-order linear differential operator, provided that we are prepared to place the problem in a non-commutative setting.

Suppose that U is an open subset of $\mathbf{R}_{p,m}$, that F is a finite-dimensional left $\mathcal{A}_{p,m}$-module, and that $f \in C^1(U, F)$. We define the *(standard) Dirac operator* D_q as

$$D_q f(x) = \sum_{j=1}^{d} q(e_j) e_j \frac{\partial f}{\partial x_j}.$$

We say that f is *Clifford analytic* if $D_q f = 0$.

First we show that this definition is independent of the choice of orthogonal basis.

Theorem 9.2.1 *Suppose that (ϵ_i) is any orthogonal basis for $\mathbf{R}_{p,m}$; denote the corresponding co-ordinates by (y_i). Then*

$$D_q = \sum_{i=1}^{d} q(\epsilon_i) \epsilon_i \frac{\partial}{\partial y_i}.$$

Proof We can express each vector ϵ_i in terms of the basis (e_j) as $\epsilon_i = \sum_{j=1}^{d} a_{ij} e_j$, and each vector e_j in terms of the basis (ϵ_i) as $e_j = \sum_{i=1}^{d} b_{ji} \epsilon_i$. Then

$$a_{ij} = b(\epsilon_i, e_j) = b(e_j, \epsilon_i) = b_{ji}.$$

Thus

$$\sum_{i=1}^{d} q(\epsilon_i)a_{ij}a_{ik} = \sum_{i=1}^{d} q(\epsilon_i)b_{ji}b_{ki} = b\left(\sum_{i=1}^{d} b_{ji}\epsilon_i, \sum_{i=1}^{d} b_{ki}\epsilon_i\right) = b(e_j, e_k).$$

Since

$$\frac{\partial f}{\partial y_i}(x) = \lim_{t\to 0} \frac{f(x + \sum_{j=1}^{d} ta_{ij}e_j) - f(x)}{t},$$

it follows that

$$\frac{\partial}{\partial y_i} = \sum_{j=1}^{d} a_{ij}\frac{\partial}{\partial x_j}.$$

Thus

$$\sum_{i=1}^{d} q(\epsilon_i)\epsilon_i\frac{\partial}{\partial y_i} = \sum_{i=1}^{d} q(\epsilon_i)\sum_{j=1}^{d} a_{ij}e_j\left(\sum_{k=1}^{d} a_{ik}\frac{\partial}{\partial x_k}\right)$$

$$= \sum_{j=1}^{d}\sum_{k=1}^{d}\left(\sum_{i=1}^{d} q(\epsilon_i)a_{ij}a_{ik}\right)e_j\frac{\partial}{\partial x_k}$$

$$= \sum_{j=1}^{d}\sum_{k=1}^{d} b(e_j, e_k)e_j\frac{\partial}{\partial x_k}$$

$$= \sum_{j=1}^{d} q(e_j)e_j\frac{\partial}{\partial x_j} = D_q.$$

□

We now consider D_q^2 as a mapping from $C^2(U, F)$ into $C(U, F)$:

$$D_q^2 = \sum_{j=1}^{d} q(e_j)e_j\frac{\partial}{\partial x_j}\left(\sum_{j=1}^{d} q(e_j)e_j\frac{\partial}{\partial x_j}\right)$$

$$= \sum_{i=1}^{d}\sum_{j=1}^{d} q(e_i)q(e_j)e_ie_j\frac{\partial}{\partial x_i}\frac{\partial}{\partial x_j}$$

$$= \sum_{i=1}^{d} q(e_i)^2 e_i^2\frac{\partial^2}{\partial x_i^2} + \sum_{i=1}^{d-1}\sum_{j=i+1}^{d} q(e_i)q(e_j)(e_ie_j + e_je_i)\frac{\partial}{\partial x_i}\frac{\partial}{\partial x_j}$$

$$= -\sum_{i=1}^{d} q(e_i)\frac{\partial^2}{\partial x_i^2} = -\Delta_q,$$

where Δ_q is the Laplacian. We have written $-\Delta_q$ as the product of two first-order differential operators; but the price that we pay is that we must work in a non-commutative setting.

Thus if we consider

$$C^2(U, F) \xrightarrow{D_q} C^1(U, F) \xrightarrow{D_q} C(U, F),$$

then

$$\ker D_q = \{f \in C^1(U, F) : f \text{ is Clifford analytic}\},$$
$$\ker D_q^2 = \{f \in C^2(U, F) : f \text{ is } q\text{-harmonic}\}.$$

Further, if f is q-harmonic, then $D_q f$ is Clifford analytic.

We shall study the Dirac operator, and Clifford analyticity, in more detail in the next chapter.

9.3 Maxwell's equations

Maxwell's equations for an electromagnetic field can be expressed as a single equation involving the standard Dirac operator.

We consider the simplest case of electric and magnetic fields in a vacuum, varying in space and time. We consider an open subset U of $\mathbf{R} \times \mathbf{R}^3 = \mathbf{R}^4$. We denote points of U by (t, x_1, x_2, x_3): t is the time variable, and x_1, x_2, x_3 are space variables. To begin with, we consider $\mathbf{E} = (E_1, E_2, E_3)$, the *electric field*, and $\mathbf{B} = (B_1, B_2, B_3)$, the *magnetic field*, as continuously differentiable vector-valued functions defined on U and taking values in $\mathbf{R}^3$. Given these, there are then a vector-valued current density, which is a continuous vector-valued function $\mathbf{J}$ defined on U and taking values in $\mathbf{R}^3$, and a charge density ρ, which is a continuous scalar-valued function ρ defined on U. Then, with a suitable choice of units, Maxwell's equations are

$$\nabla.\mathbf{E} = \rho,$$
$$\nabla \times \mathbf{B} - \frac{\partial \mathbf{E}}{\partial t} = \mathbf{J},$$
$$\frac{\partial \mathbf{B}}{\partial t} + \nabla \times \mathbf{E} = 0,$$
$$\nabla.\mathbf{B} = 0.$$

The formulation of these equations is one of the major achievements of nineteenth-century physics. It turned out however that these equations are not invariant under Euclidean changes of co-ordinates; instead, they are invariant under Lorentz transformations. This was one of the driving forces that led to Einstein's theory of special relativity. Thus, instead of $\mathbf{R}^4$ we must consider $\mathbf{R}_{1,3}$, with quadratic form

$$q(t, x_1, x_2, x_3) = t^2 - (x_1^2 + x_2^2 + x_3^2)$$

and orthogonal basis (e_0, e_1, e_2, e_3). (It would also be possible to work with $\mathbf{R}_{3,1}$.) We therefore consider U to be an open subset of $\mathbf{R}_{1,3}$.

We must also consider the vector spaces which contain the images of E, B and J more carefully. We now consider the electric and magnetic fields as continuously differentiable mappings from U into the Clifford algebra $\mathcal{A}_{1,3}$. But instead of being vector-valued, we require them to be bivector-valued. The vector space V of bivectors is a six-dimensional hyperbolic space: we can write it as $V = V^+ \oplus V^-$, where

$$V^+ = \text{span}(e_2 e_3, e_1 e_3, e_1 e_3)$$
$$V^- = \text{span}(e_0 e_1, e_0 e_2, e_0 e_1);$$

the quadratic form Q on $\mathcal{A}_{1,3}$ is positive definite on V^+ and is negative definite on V^-.

In order to avoid confusion, we shall write

$$\boldsymbol{E} = E_1 e_1 + E_2 e_2 + E_3 e_3 \text{ and } \boldsymbol{B} = B_1 e_1 + B_2 e_2 + B_3 e_3$$

as the original $\mathbf{R}^3$-valued functions. We set

$$\tilde{\boldsymbol{E}} = E_1 e_0 e_1 + E_2 e_0 e_2 + E_3 e_0 e_3 = e_0 \boldsymbol{E},$$
$$\tilde{\boldsymbol{B}} = B_1 e_2 e_3 - B_2 e_1 e_3 + B_3 e_1 e_2 = \boldsymbol{B} e_1 e_2 e_3.$$

Thus $\tilde{\boldsymbol{B}}$ takes values in V^+ and $\tilde{\boldsymbol{E}}$ takes values in V^-. We set $\tilde{\boldsymbol{F}} = \tilde{\boldsymbol{E}} + \tilde{\boldsymbol{B}}$; $\tilde{\boldsymbol{E}}$, $\tilde{\boldsymbol{B}}$ and $\tilde{\boldsymbol{F}}$ are all bivector-valued.

We combine the current density and the charge density: we set $\tilde{\boldsymbol{J}} = \rho e_0 + \boldsymbol{J}$, so that $\tilde{\boldsymbol{J}}$ is vector-valued, taking values in the four-dimensional space $\mathbf{R}_{1,3}$.

We now calculate $D_q \tilde{\boldsymbol{F}}$. We can write $D_q \tilde{\boldsymbol{F}} = (D_q \tilde{\boldsymbol{F}})_1 + (D_q \tilde{\boldsymbol{F}})_3$, where $(D_q \tilde{\boldsymbol{F}})_1$ takes values in $\mathbf{R}_{1,3}$, and is the vector part of $D_q \tilde{\boldsymbol{F}}$, and $(D_q \tilde{\boldsymbol{F}})_3$ takes values in $\mathbf{R}_{1,3} e_\Omega$, and is the trivector part of $D_q \tilde{\boldsymbol{F}}$. First,

$$(D_q \tilde{\boldsymbol{F}})_1 = -\sum_{j=1}^{3} \frac{\partial E_j}{\partial t} e_j - \sum_{j=1}^{3} \frac{\partial E_j}{\partial x_j} e_0 + \left(\frac{\partial B_2}{\partial x_3} - \frac{\partial B_3}{\partial x_2} \right) e_1$$
$$+ \left(\frac{\partial B_3}{\partial x_1} - \frac{\partial B_1}{\partial x_3} \right) e_2 + \left(\frac{\partial B_1}{\partial x_2} - \frac{\partial B_2}{\partial x_1} \right) e_3$$
$$= \rho e_0 + \boldsymbol{J} = \tilde{\boldsymbol{J}},$$

and secondly

$$(D_q \tilde{F})_3 = T_1 + T_2 + T_3,$$

where

$$T_1 = \left(\frac{\partial E_2}{\partial x_3} - \frac{\partial E_3}{\partial x_2} \right) e_0 e_2 e_3 + \left(\frac{\partial E_3}{\partial x_1} - \frac{\partial E_1}{\partial x_3} \right) e_0 e_1 e_3$$

$$+ \left(\frac{\partial E_1}{\partial x_2} - \frac{\partial E_2}{\partial x_1} \right) e_0 e_1 e_2,$$

$$T_2 = \frac{\partial B_1}{\partial t} e_0 e_2 e_3 - \frac{\partial B_2}{\partial t} e_0 e_1 e_3 + \frac{\partial B_3}{\partial t} e_0 e_1 e_2$$

$$T_3 = \frac{\partial B_1}{\partial x_1} e_1 e_2 e_3 - \frac{\partial B_2}{\partial x_2} e_2 e_1 e_3 + \frac{\partial B_3}{\partial x_3} e_3 e_1 e_2.$$

Thus

$$(D_q \tilde{F})_3 = -(\nabla \times E) e_\Omega - \left(\frac{\partial B}{\partial t} \right) e_\Omega + (\nabla . B) e_1 e_2 e_3 = 0;$$

Hence Maxwell's equations reduce to the single equation $D_q \tilde{F} = \tilde{J}$. Note that this implies that if $\tilde{J} = 0$ then $\tilde{F}$ is Clifford analytic, and that if $\tilde{J}$ is Clifford analytic then E and B are harmonic.

9.4 The Dirac equation

Our next example concerns the Dirac equation. This relativistic quantum mechanical equation was defined by Paul Dirac in 1928; it led to the discovery of the positron, and is one of the great landmarks of theoretical physics. It was here that the Dirac operator was introduced.

We begin with the classical relativistic equation

$$E^2 = k.k + m^2,$$

where, with suitable units, E is energy, k is momentum and m is rest-mass. The corresponding equation in quantum mechanics is the Klein-Gordon equation. The underlying Hilbert space H is a suitable space of sufficiently smooth functions, called *wave functions*, on $\mathbf{R}^4$. In this quantum mechanical setting,

E^2 corresponds to $\dfrac{\partial^2}{\partial t^2}$ and $k.k$ corresponds to $\dfrac{\partial^2}{\partial x_1^2} + \dfrac{\partial^2}{\partial x_2^2} + \dfrac{\partial^2}{\partial x_3^2}.$

Thus we are led to the *Klein-Gordon equation*

$$\Box^2\psi = \frac{\partial^2\psi}{\partial t^2} - \left(\frac{\partial^2\psi}{\partial x_1^2} + \frac{\partial^2\psi}{\partial x_2^2} + \frac{\partial^2\psi}{\partial x_3^2}\right) = -m\psi^2,$$

where ψ is a wave function.

There are two conventions that we can consider for space-time: either $\mathbf{R}_{3,1}$ or $\mathbf{R}_{1,3}$. We shall consider each of them, and begin with $\mathbf{R}_{3,1}$. We therefore consider a wave function ψ taking values in a left $\mathcal{A}_{3,1}$ module F. If $D_{3,1}$ is the corresponding Dirac operator, then $D_{3,1}^2 = \Box^2$. Consequently, we consider the equation

$$D_{3,1}\psi = im\psi, \quad \text{where } i \text{ is an operator with } i^2 = -1.$$

The operator i must commute with e_1, e_2, e_3 and e_4. But $\mathcal{A}_{3,1} \cong M_2(\mathbf{H})$, and $Z(\mathcal{A}_{3,1}) \cong \mathbf{R}$. We therefore need to ensure that F is also a $\mathbf{C}$-module; that is, we require that F is a *complex* vector space.

We follow the path that Dirac took. As in Section 7.6, we can consider $\mathcal{A}_{3,1}$ as a subalgebra of $\mathcal{A}_{3,2}$, and $\mathcal{A}_{3,2}$ is isomorphic to $M_4(\mathbf{C})$. A spinor space for $\mathcal{A}_{3,2}$ is isomorphic to $\mathbf{C}^4$, and so we consider $\mathcal{A}_{3,1}$ acting on $\mathbf{C}^4$. For this reason, we consider the wave function ψ taking values in $\mathbf{C}^4$. We can use the Dirac γ-matrices defined in Section 7.6 to represent $\mathcal{A}_{3,1}$ in $M_4(\mathbf{C})$; the Dirac equation then becomes

$$D_{3,1}\psi = -\gamma_0\frac{\partial\psi}{\partial t} + \gamma_1\frac{\partial\psi}{\partial x_1} + \gamma_2\frac{\partial\psi}{\partial x_2} + \gamma_3\frac{\partial\psi}{\partial x_3} = im\psi.$$

Alternatively, we can consider space-time as $\mathbf{R}_{1,3}$, and label the standard orthogonal basis as (e_0, e_1, e_2, e_3). In this case, $D_{1,3}^2 = -\Box^2$, and so we consider the equation $D_{1,3}\psi = m\psi$. Now $\mathcal{A}_{1,3} \cong M_4(\mathbf{R})$, and, as in Section 7.6, but with a change of indices, we obtain a representation by using the matrices

$$j_0 = J \otimes U, \quad j_1 = Q \otimes U, \quad j_2 = U \otimes U, \quad j_3 = I \otimes Q.$$

A spinor space for $\mathcal{A}_{1,3}$ is isomorphic to $\mathbf{R}^4$, and we could consider ψ taking values in $\mathbf{R}^4$. In quantum mechanics, however, wave functions are required to be complex; we therefore take ψ to take values in $\mathbf{C}^4$. The Dirac equation then becomes

$$D_{1,3}\psi = j_0\frac{\partial\psi}{\partial t} - j_1\frac{\partial\psi}{\partial x_1} - j_2\frac{\partial\psi}{\partial x_2} - j_3\frac{\partial\psi}{\partial x_3} = m\psi.$$

But in this case, the matrices are real-valued, with entries taking values -1, 0 and 1.

Exercises

1. Let $\psi = (\psi_1, \psi_2, \psi_3, \psi_4)$ take values in $\mathbf{C}^4$. Expand the Dirac equation, using the matrices $\gamma_1, \gamma_2, \gamma_3, \gamma_4$ defined above, to obtain four first-order linear differential equations.

2. Do the same, where now $\psi = (\psi_1, \psi_2, \psi_3, \psi_4)$ take values in $\mathbf{R}^4$, using the matrices j_0, j_1, j_2, j_3 defined above.

10

Clifford analyticity

We now study the Dirac operator, and Clifford analyticity, in greater detail. We can identify the complex field $\mathbf{C}$ with two-dimensional real space $\mathbf{R}^2$ and can compare analytic functions on $\mathbf{C}$ with harmonic functions on $\mathbf{R}^2$. The relations between these two notions has led to a great deal of profound analysis. What happens in higher dimensions? There are corresponding relations, of a weaker, dimensionally dependent, sort between harmonic functions and Clifford analytic functions, and we shall describe some of these in this chapter. Although we shall describe the two-dimensional background, it will help to have some familiarity with the relationship between analytic and harmonic functions in the plane.

10.1 Clifford analyticity

Let us recall the definitions. Suppose that U is an open subset of $\mathbf{R}_{p,m}$, that F is a finite-dimensional left $\mathcal{A}_{p,m}$-module, and that $f \in C^1(U, F)$. Then the *(standard) Dirac operator* D_q is defined as

$$D_q f(x) = \sum_{j=1}^{d} q(e_j) e_j \frac{\partial f}{\partial x_j},$$

and f is *Clifford analytic* if $D_q f = 0$.

Further, $D_q^2 = -\Delta_q$, so that if f is Clifford analytic, then f is q-harmonic, and if $f \in C^2(U, F)$ is q-harmonic, then $D_q(f)$ is Clifford analytic.

Let us first give some examples of q-harmonic functions; the examples in Euclidean space generalize to spaces with other quadratic forms.

Proposition 10.1.1 *If $q(x) > 0$, let $f(x) = \log q(x)$ for $d = 2$ and* let $f(x) = 1/q(x)^{(d-2)/2}$ for $d > 2$. Then f is q-harmonic on the set $\{x \in \mathbf{R}_{p,m} : q(x) > 0\}$.

Proof If $d = 2$ then

$$\frac{\partial f}{\partial x_j} = \frac{2q(e_j)x_j}{q(x)} \quad \text{and} \quad \frac{\partial^2 f}{\partial x_j^2} = \frac{2q(e_j)}{q(x)} - \frac{4x_j^2}{q(x)^2},$$

so that

$$\Delta_q f(x) = q(e_1)\frac{\partial^2 f}{\partial x_1^2} + q(e_2)\frac{\partial^2 f}{\partial x_2^2} = \frac{4}{q(x)} - \frac{4(q(e_1)x_1^2 + q(e_2)x_2^2)}{q(x)^2} = 0.$$

If $d > 2$ then

$$\frac{\partial f}{\partial x_j} = \frac{-(d-2)q(e_j)x_j}{q(x)^{d/2}},$$

so that

$$\frac{\partial^2 f}{\partial x_j^2} = \frac{-(d-2)q(e_j)}{q(x)^{d/2}} + \frac{d(d-2)x_j^2}{q(x)^{(d+2)/2}},$$

so that

$$\Delta_q f(x) = \sum_{j-1}^{d} q(e_j)\frac{\partial^2 f}{\partial x_j^2} = \frac{-d(d-2)}{q(x)^{d/2}} + \frac{d(d-2)q(x)}{q(x)^{(d+2)/2}} = 0.$$

$\square$

If f is a real-valued function on $\mathbf{R}_{p,m}$, we can consider it as a function taking values in the scalars in $\mathcal{A}_{p,m}$.

Corollary 10.1.1 *The $\mathbf{R}_{p,m}$-valued function $g(x) = x/q(x)^{d/2}$ is Clifford analytic on $\{x : q(x) > 0\}$.*

Proof If $d = 2$ then $D_q(\log q(x)) = 2x/q(x)$ and if $d > 2$ then

$$D_q(q(x)^{-(d-2)/2}) = -(d-2)x/q(x)^{d/2}.$$

$\square$

Let us consider the case where $F = \mathcal{A}_{p,m}$ and where $f = \sum_{i=1}^{n} f_i e_i$ takes values in $\mathbf{R}_{p,m} \subseteq \mathcal{A}_{p,m}$. In this case,

$$D_q f(x) = \sum_{i=1}^{d} \left(\sum_{j=1}^{d} q(e_j)e_j \frac{\partial f_i}{\partial x_j} e_i \right)$$

$$= -\sum_{i=1}^{d} q(e_i)\frac{\partial f_i}{\partial x_i} + \sum_{1 \le i < j \le d} \left(q(e_i)\frac{\partial f_j}{\partial x_i} - q(e_j)\frac{\partial f_i}{\partial x_j} \right) e_i e_j.$$

Thus in this case f is Clifford analytic if and only if it satisfies the *generalized Cauchy-Riemann* equations (the *GCR equations*):

$$\sum_{i=1}^{d} q(e_i) \frac{\partial f_i}{\partial x_i} = 0,$$

$$q(e_i) \frac{\partial f_j}{\partial x_i} = q(e_j) \frac{\partial f_i}{\partial x_j} \text{ for } i \neq j.$$

As another example, let us consider functions taking values in $\mathbf{R}_3$. In $\mathbf{R}_3$, $\mathrm{curl}(f) = \nabla \times f$ is defined as

$$\nabla \times f = \left(\frac{\partial f_2}{\partial x_3} - \frac{\partial f_3}{\partial x_2} \right) e_1 + \left(\frac{\partial f_3}{\partial x_1} - \frac{\partial f_1}{\partial x_3} \right) e_2 + \left(\frac{\partial f_1}{\partial x_2} - \frac{\partial f_1}{\partial x_2} \right) e_3.$$

In this case, then,

$$D_q(f) = -\nabla.f - (\nabla \times f) e_\Omega.$$

There are certain situations where we can consider the Dirac operator as a graded operator. First, suppose that $F = \mathcal{A}_{p,m} = \mathcal{A}$. Then

$$D_q(C^1(U, \mathcal{A}^+)) \subseteq C(U, \mathcal{A}^-) \text{ and } D_q(C^1(U, \mathcal{A}^-)) \subseteq C(U, \mathcal{A}^+).$$

Secondly, suppose that F is a spinor space for $\mathcal{A}_{p,m}$. Then it may or may not happen that the action of $\mathcal{A}_{p,m}^+$ on F is irreducible. For example, if F is a spinor space for $\mathcal{A}_6$, then $\dim F = 8$. Since $\mathcal{A}_6^+ \cong \mathcal{A}_5$, and spinor spaces for $\mathcal{A}_5$ also have real dimension 8, the action of $\mathcal{A}_6^+$ on F is irreducible. On the other hand, if F is a spinor space for $\mathcal{A}_{3,1}$ then F has real dimension 8, whereas a spinor space for $\mathcal{A}_{2,1}$ has real dimension 4, and so $F = F_1 \oplus F_2$ is the direct sum of two $\mathcal{A}_{3,1}^+$-invariant subspaces, and

$$D_q(C^\infty(U, F_1)) \subseteq C^\infty(U, F_2) \text{ and } D_q(C^\infty(U, F_2)) \subseteq C^\infty(U, F_1).$$

10.2 Cauchy's integral formula

Cauchy's integral formula is one of the most powerful tools in complex analysis. This section is descriptive; it describes how the Cauchy integral formula extends to Clifford analytic functions. We restrict our attention to Euclidean space. We use the language and terminology of differential geometry. We begin by expressing the Dirac operator in terms of the exterior derivative of a vector-valued $(d-1)$-form. Let U be an open subset of the Euclidean space $\mathbf{R}_d$. Let dV denote the volume element d-form $dV = dx_1 \wedge \cdots \wedge dx_d$, and for $1 \le j \le d$, let $d\hat{x}_j$ be the $(d-1)$-form

$$d\hat{x}_j = dx_1 \wedge \cdots \wedge \widehat{dx_j} \wedge \cdots dx_d$$

(where the term dx_j is omitted), so that $dV = (-1)^{j-1}dx_j \wedge d\hat{x}_j$. Now let $d\sigma$ be the vector-valued $(d-1)$-form

$$d\sigma = \sum_{j=1}^{d}(-1)^{j-1}e_j d\hat{x}_j.$$

Then if f is a smooth function on U taking values in a left $\mathcal{A}_d$-module F there exists an F-valued $(d-1)$-form

$$\omega_f(x) = d\sigma f(x) = \sum_{j=1}^{d}(-1)^{j-1}e_j f(x)d\hat{x}_j.$$

(Note the order of the terms.) This has exterior derivative

$$dw_f(x) = \sum_{i=1}^{d} dx_i \wedge \frac{\partial w_f}{\partial x_i}(x)$$

$$= \sum_{i=1}^{d} dx_i \wedge \left(\sum_{j=1}^{d}(-1)^{j-1}e_j \frac{\partial f}{\partial x_i}(x)d\hat{x}_j \right)$$

$$= \sum_{i=1}^{d} e_i \frac{\partial f}{\partial x_i}(x)dV = D_q f(x)dV.$$

As a result of this, it can be shown that if f is Clifford analytic, then the following analogue of Cauchy's integral formula holds: if B is a closed ball contained in U and y is an interior point of B then

$$f(y) = -\frac{1}{A_{d-1}} \int_{\partial B} \frac{x-y}{\|x-y\|^d} d\sigma f(x),$$

where A_{d-1} is the $d-1$ volume of the unit sphere in $\mathbf{R}_d$.

10.3 Poisson kernels and the Dirichlet problem

In this section we give a brief descriptive account of some of the fundamental ideas and results from harmonic analysis in Euclidean space. Let $H^{d+1} = \{(t,x) \in \mathbf{R} \times \mathbf{R}^d : t > 0\}$ denote the upper half-space in $\mathbf{R}^{d+1}$, and let λ denote Lebesgue measure on $\mathbf{R}^d$.

The *Poisson kernel*

$$P(t,x) = P_t(x) = \frac{c_d t}{(x^2 + t^2)^{(d+1)/2}},$$

(where c_d is a normalizing constant, chosen so that $\int_{\mathbf{R}^d} P_t(x)\, d\lambda(x) = 1$) is a harmonic function on H^{d+1} (see the exercises below).

Suppose that ν is a complex Borel measure on $\mathbf{R}^d$. We can define a function u on the upper half-space H^{d+1} by setting

$$u(t,x) = P(\nu)(t,x) = P_t(\nu)(x) = \int_{\mathbf{R}} P_t(x-y)\, d\nu(y).$$

Then u is harmonic on $\mathbf{H}^{d+1}$.

Similarly, if $1 \le p < \infty$ and $f \in L^p(\lambda)$, we set

$$P(f)(t,x) = P_t(f)(x) = \int_{\mathbf{R}^d} P_t(x-y)f(y)\, d\lambda(y).$$

$P(f)$ is again harmonic. Then the Poisson kernel is an approximate identity, which solves the Dirichlet problem for the upper half-space.

Theorem 10.3.1 *If $f \in L^p(\lambda)$ then $P(f)$ is harmonic on the upper half-space H^{d+1}. Further, $P_t(f) \to f$ in L^p-norm, and almost everywhere, and in particular at the points of continuity of f, as $t \searrow 0$.*

Instead of starting with measures and functions on $\mathbf{R}^d$, we can start with harmonic functions on H^{d+1}. This raises the following fundamental question. Suppose that g is a harmonic function on the upper half-space H^{d+1}; let $g_t(x) = g(t,x)$. For $1 \le p < \infty$, we define the space $h_p(H^{d+1})$ to be the space of harmonic functions g on H^{d+1} for which

$$\|g\|_{h_p} = \sup\{\|g_t\|_p : t > 0\} < \infty.$$

Then $(h_p(H^{d+1}), \|g\|_{h_p})$ is a Banach space. If $g \in h_p(H^{d+1})$, does g_t converge to an element of $L^p(\lambda)$ as $t \searrow 0$, and if so, does it converge pointwise almost everywhere?

When $1 < p < \infty$ then the answer is 'yes'. The mapping $f \to P(f)$ is a linear isomorphism of $L^p(\lambda)$ onto $h_p(H^{d+1})$.

When $p = 1$, the answer is 'no'. In fact, if $M(\mathbf{R}^d)$ is the linear space of complex Borel measures on $\mathbf{R}^d$, then the mapping $\nu \to P(\nu)$ is a linear isomorphism of $M(\mathbf{R}^d)$ onto $h_1(H^{d+1})$. $L^1(\lambda)$ is a closed linear subspace of $M(\mathbf{R}^d)$, and the image $P(L^1(\lambda))$ is a proper linear subspace of $h_1(H^{d+1})$.

Let us introduce some spaces which will be of interest later. Suppose that g is a function on the upper half-space H^{d+1}. We define the *maximal* *function g^* on $\mathbf{R}^d$* by setting $g^*(x) = \sup_{t>0} |g(t,x)|$. For $1 \le p < \infty$ we define the *Hardy space $H_p(H^{d+1})$* to be

$$H_p(H^{d+1}) = \{g \in h_p(H^{d+1}) : g^* \in L^p(\lambda)\}.$$

Similarly, we define the *Hardy space $H_p(\mathbf{R}^d)$* to be

$$\{f \in L^p(\lambda) : P(f) \in H_p(H^{d+1})\}.$$

We then have the following result.

Theorem 10.3.2 *If $1 < p < \infty$ then $H_p(\mathbf{R}^d) = L^p(\mathbf{R}^d)$. On the other hand, $H_1(\mathbf{R}^d)$ is a proper linear subspace of $L^1(\mathbf{R}^d)$. If $g \in H_1(H^{d+1})$ then there exists $f \in H^1(\mathbf{R}^d)$ such that $P(f) = g$. Thus $P(H_1(\mathbf{R}^d)) = H_1(H^{d+1})$.*

If χ_B is the characteristic function of the unit ball in $\mathbf{R}^d$, then $\chi_B \in L^1(\mathbf{R}^d) \setminus H_1(\mathbf{R}^d)$.

Exercises

1. If $z = x + it \in \mathbf{C}$, show that
$$\frac{i}{\pi z} = \frac{t}{\pi(x^2 + t^2)} + \frac{ix}{\pi(x^2 + t^2)}$$
$$= P(x,t) + iQ(x,t) = P_t(x) + iQ_t(x).$$

P is the Poisson kernel, and Q is the *conjugate Poisson kernel* on $\mathbf{R}$.
2. Deduce that P and Q are harmonic functions on H^2.
3. Verify that $\int_R P_t(x)\, dx = 1$.
4. Verify that the function $P(t,x) = c_d t(x^2 + t^2)^{-(d+1)/2}$ is harmonic on the upper half-space H^{d+1}.

10.4 The Hilbert transform

We now consider what happens in the case where $d = 1$. We can identify H^2 with the upper half-plane $\mathbf{C}^+ = \{z = x + it \in \mathbf{C} : t > 0\}$. If u is harmonic on $\mathbf{C}^+$ then u is the real part of an analytic function $u + iv$ on $\mathbf{C}^+$, which is unique up to an additive constant.

We can use the conjugate Poisson kernel Q (defined in the exercises in the previous section) to define $Q(\nu)$ and $Q(f)$, exactly as above. Then $P(\nu) + iQ(\nu)$ is an analytic function on $\mathbf{C}^+$, as is $P(f) + iQ(f)$.

When $1 < p < \infty$, things continue to work well. If $f \in L^p(\lambda)$ then

$Q(f) \in h_p(H^2)$, so that $Q_t(f)$ converges in L^p-norm to a function $H(f)$ as $t \searrow 0$. $H(f)$ is the *Hilbert transform* of f, and the mapping $f \to H(f)$ is a linear homeomorphism of $L^p(\lambda)$ onto itself. When $p = 2$, the Hilbert transform is in fact an isometry.

When $p = 1$, things do not work so well, since $Q_t(L^1)$ is not contained in L^1. We must consider $H_1(\mathbf{R})$, rather than L^1. If $f \in H_1(\mathbf{R})$, then $Q(f) \in H_1(H^2)$, so that $Q_t(f)$ converges in L^1-norm to a function $H(f)$ as $t \searrow 0$. Then $H(f) \in H_1(\mathbf{R})$, and the Hilbert transform is a linear homeomorphism of $H_1(\mathbf{R})$ onto itself.

The following remarkable theorem, due to the brothers Riesz, shows that this is the only sensible course to take.

Theorem 10.4.1 (F. and M. Riesz)   *Suppose that ν is a bounded complex measure on $\mathbf{R}$ for which $P(\nu)$ is analytic on $\mathbf{C}^+$. Then there exists $f \in H_1(\mathbf{R})$ such that $\nu = f \, d\lambda$.*

These results show that the relation between harmonic functions and complex functions is subtle, and is not straightforward. Our aim will be to see how Clifford algebras and augmented Dirac operators can be used to extend these results to higher dimensions.

Exercise

1. Let f be the indicator function of $[-1, 1]$. Calculate $Q_t(f)(x)$ for $|x| > 1$, and show that $Q_t(f)$ is not in $L^1(\mathbf{R})$.

10.5 Augmented Dirac operators

Suppose that ν is a bounded signed or complex measure on $\mathbf{R}^d$. Using the d-dimensional *Poisson kernel*, we have defined a function $P(\nu)$ on the upper half-space $H^{d+1} = \{(x, t) \in \mathbf{R}^{d+1} : t > 0\}$. The new variable t is separate from the original variable x. Putting this in a Clifford algebra context, the function u can be considered as a function on a subset of $\mathbf{R} \oplus \mathbf{R}^d \subseteq \mathcal{A}_d$. (More generally, an element of the subspace $\mathbf{R} \oplus \mathbf{R}_{p,m}$ of $\mathcal{A}_{p,m}$ is called a *paravector*.) We denote such a paravector by $(t, x_1, \ldots, x_d)$.

This means that we need to consider a differential operator more general than the Dirac operator that we have considered in the previous chapter. Suppose that U is an open subset of the space $\mathbf{R} \oplus \mathbf{R}_{p,m}$ of paravectors in $\mathcal{A}_{p,m}$, that F is a finite-dimensional left $\mathcal{A}_{p,m}$-module, and that $f : U \to F$ is a smooth function. We define the *augmented Dirac operator* $\mathcal{D}$ to be

$$\mathcal{D}f = \frac{\partial f}{\partial t} + \sum_{j=1}^{d} q(e_j)e_j \frac{\partial f}{\partial x_j},$$

and its conjugate $\overline{\mathcal{D}}$, to be

$$\overline{\mathcal{D}}f = \frac{\partial f}{\partial t} + \sum_{j=1}^{d} q(e_j)\bar{e}_j \frac{\partial f}{\partial x_j} = \frac{\partial f}{\partial t} - \sum_{j=1}^{d} q(e_j)e_j \frac{\partial f}{\partial x_j}.$$

Then

$$\mathcal{D}\overline{\mathcal{D}}(f) = \overline{\mathcal{D}}\mathcal{D}(f) = \frac{\partial^2 f}{\partial t^2} + \sum_{i=1}^{d} q(e_i)\frac{\partial^2 f}{\partial x_i^2} = \Delta,$$

where now Δ is the Laplacian on $\mathbf{R} \oplus \mathbf{R}_{p,m} \cong \mathbf{R}_{p+1,m}$.

We say that f is an (augmented) Clifford analytic function if $\mathcal{D}(f) = 0$. In particular, if f is harmonic on U then $\overline{\mathcal{D}}f$ is analytic.

Frequently, one considers functions

$$u = (u_0, u_1, \ldots, u_n) = u_0 + u_1 e_1 + \cdots + u_n e_n$$

taking values in $\mathbf{R} \oplus \mathbf{R}_{p,m} \subseteq \mathcal{A}_{p,m}$. Then the condition $\mathcal{D}f = 0$ reduces to the *augmented Cauchy-Riemann* (ACR) equations:

$$\frac{\partial u_0}{\partial t} = \sum_{i=1}^{d} q(e_i)\frac{\partial u_i}{\partial x_i}$$

$$q(e_j)\frac{\partial u_0}{\partial x_j} = -\frac{\partial u_j}{\partial t} \qquad 1 \le j \le d$$

$$q(e_j)\frac{\partial u_i}{\partial x_j} = q(e_i)\frac{\partial u_j}{\partial x_i} \qquad 1 \le i < j \le d.$$

Let us see how the augmented Cauchy-Riemann equations relate to the familiar Cauchy-Riemann equations of complex analysis. Suppose that U is an open subset of $\mathbf{R} \oplus \mathbf{R}$ and that f is a smooth mapping of U into $\mathcal{A}_1 \cong \mathbf{C}$, and that $f(x,y) = u(x,y) \oplus v(x,y)e_1$. We set

$$\frac{d}{dz} = \frac{1}{2}\left(\frac{\partial}{\partial x} - e_1\frac{\partial}{\partial y}\right) = \frac{1}{2}\overline{\mathcal{D}}$$

$$\frac{d}{d\bar{z}} = \frac{1}{2}\left(\frac{\partial}{\partial x} + e_1\frac{\partial}{\partial y}\right) = \frac{1}{2}\mathcal{D}.$$

Then f is analytic if $df/d\bar{z} = 0$ and if f is analytic then its derivative is df/dz. The equation $df/d\bar{z} = 0$ expands to give the Cauchy-Riemann equations.

The fact that d/dz relates to $\overline{\mathcal{D}}$ and $d/d\bar{z}$ relates to $\mathcal{D}$, rather than the other way round, is unfortunate. It would be appropriate to change the signs in the definition of $\mathcal{D}$, so that

$$\mathcal{D}f = \frac{\partial f}{\partial t} + \sum_{j=1}^{d} e_j^{-1}\frac{\partial f}{\partial x_j},$$

but the definitions that we have given conform to the usual practice.

10.6 Subharmonicity properties

Why are analytic functions in the plane better behaved than harmonic functions? Many of the deep results concerning analytic functions depend upon their subharmonicity properties.

A measurable real-valued function f on an open subset W of $\mathbf{R}_d$, or $\mathbf{C}$, is said to be *subharmonic* if, whenever U is a ball contained in W with centre x, then

$$f(x) \leq \frac{1}{\lambda(U)} \int_U f(y)\,d\lambda(y).$$

If f is smooth then f is subharmonic if and only if $\Delta(f) \geq 0$. If f is an analytic function on an open subset U of $\mathbf{C}$ then $\log|f|$ is subharmonic, and so $|f|^\alpha$ is subharmonic, for $0 < \alpha < \infty$. Many results concerning analytic functions on an open subset U of $\mathbf{C}$ follow from these facts. As we shall now see, there are corresponding results for Clifford analytic functions, but they are less strong, and depend upon the dimension d. We restrict attention to the Euclidean case.

Theorem 10.6.1 *Suppose that f is a Clifford analytic function on an open subset U of $\mathbf{R}_d$, taking non-zero values in $\mathbf{R}_d \subseteq \mathcal{A}_d$. Then $\Delta(\|f\|^q) \geq 0$, for $q \geq (d-2)/(d-1)$.*

Proof We need the following lemma.

Lemma 10.6.1 *Suppose that $M = (m_{jk}) \in M_d(\mathbf{R})$ is symmetric and has trace 0. Then*

$$\|M\|^2 \leq \frac{d-1}{d}\sum_{j,k} m_{jk}^2,$$

where $\|M\|$ is the operator norm of M acting on l_2^d.

Proof We diagonalize M: there exists an orthogonal matrix P such that $P^\star M P = \text{diag}(\lambda_1, \ldots, \lambda_d)$. Then $\|M\| = \max_j |\lambda_j| = |\lambda_1|$, say,

$$\sum_{j,k} m_{jk}^2 = \tau(M^\star M) = \sum_{j=1}^{d} \lambda_j^2,$$

and $\tau(M) = \sum_{j=1}^{d} \lambda_j = 0$. Thus

$$\lambda_1^2 = \left(\sum_{j=2}^{d} \lambda_j \right)^2 \leq (d-1) \sum_{j=2}^{d} \lambda_j^2,$$

by the Cauchy-Schwarz inequality, so that

$$d \|M\|^2 = d\lambda_1^2 = \lambda_1^2 + (d-1)\lambda_1^2 \leq (d-1) \sum_{j=1}^{d} \lambda_j^2 = (d-1) \sum_{j,k} m_{jk}^2.$$

$\square$

We now prove Theorem 10.6.1. Since f is harmonic, the result is true if $q \geq 1$. We can therefore suppose that $(d-2)/(d-1) \leq q < 1$ (this is the interesting case).
$\|f\|^q = \langle f, f \rangle^{q/2}$, so that

$$\frac{\partial}{\partial x_j}(\|f\|^q) = q \langle f, f \rangle^{(q-2)/2} \left\langle \frac{\partial f}{\partial x_j}, f \right\rangle$$

and

$$\frac{\partial^2}{\partial x_j^2}(\|f\|^q) = q(q-2) \langle f, f \rangle^{(q-4)/2} \left\langle \frac{\partial f}{\partial x_j}, f \right\rangle^2$$
$$+ q \langle f, f \rangle^{(q-2)/2} \left(\left\langle \frac{\partial f}{\partial x_j}, \frac{\partial f}{\partial x_j} \right\rangle + \left\langle \frac{\partial^2 f}{\partial x_j^2}, f \right\rangle \right)$$

Adding, and using the harmonicity of f,

$$\Delta(\|f\|^q) = q \|f\|^{q-4} \left((q-2) \sum_j \left\langle \frac{\partial f}{\partial x_j}, f \right\rangle^2 + \|f\|^2 \sum_{k,j} \left(\frac{\partial f_k}{\partial x_j} \right)^2 \right).$$

Now

$$\sum_j \left\langle \frac{\partial f}{\partial x_j}, f \right\rangle^2 = \sum_j \left(\sum_k \frac{\partial f_k}{\partial x_j} f_k \right)^2 = \|M(f)\|^2,$$

where $M_{jk} = \partial f_k / \partial x_j$. But the generalized Cauchy-Riemann equations of Section 9.2 show that M is symmetric and that $\tau(M) = 0$. Thus by Lemma 10.6.1

$$\sum_j \left\langle \frac{\partial f}{\partial x_j}, f \right\rangle^2 \leq \frac{d-1}{d} \left(\sum_{k,j} \left(\frac{\partial f_k}{\partial x_j} \right)^2 \right) \|f\|^2,$$

and so, since $q < 2$,

$$\Delta(|f|^q) \geq q \|f\|^{q-2} \left(\frac{d-1}{d}(q-2) + 1 \right) \sum_{k,j} \left(\frac{\partial f_k}{\partial x_j} \right)^2$$

$$= \frac{q(d-1)}{d} \|f\|^{q-2} \left(q - \frac{d-2}{d-1} \right) \sum_{k,j} \left(\frac{\partial f_k}{\partial x_j} \right)^2 \geq 0.$$

$\square$

We have a corresponding result for augmented Clifford analytic functions.

Theorem 10.6.2 *Suppose that f is an augmented Clifford analytic function defined on an open subset of $\mathbf{R} \oplus \mathbf{R}_d \subseteq \mathcal{A}_d$. Then $\Delta(|f|^q) \geq 0$ for $q \geq (d-1)/d$.*

Proof Let $V = \{(s,x) \in R_{d+1} : (-s,x) \in U\}$, and let $\pi(s,x) = (-s,x)$ for $(s,x) \in V$. Then $f \circ \pi$ satisfies the generalized Cauchy-Riemann equations on V, and so $\Delta(|f \circ \pi|^q) \geq 0$ for $q \geq (d-1)/d$. The result follows, since $\Delta(|f|^q) \circ \pi = \Delta(|f \circ \pi|^q)$. $\square$

Exercise

1. Let f be the Clifford analytic function $x/\|x\|^q$ on $\mathbf{R}_d \setminus \{0\}$. Show that $|f|^q$ is not subharmonic for $0 < q < (d-2)/(d-1)$.

10.7 The Riesz transform

We are now in a position to see how results concerning the Hilbert transform can be extended to higher dimensions. Suppose that $d \geq 2$. We consider the half-space H^{d+1} as a subspace of the space $\mathbf{R} \oplus \mathbf{R}^d$ of paravectors in $\mathcal{A}_d$. By Proposition 10.1.1,

$$g(t,x) = \frac{1}{(t^2 + \|x\|^2)^{(d-1)/2}}$$

is harmonic on $(\mathbf{R} \times \mathbf{R}^d) \setminus \{(0,0)\}$.

Let $h(t,x) = -c_d g(t,x)/(d-1)$, where c_d is a normalizing constant. For $t > 0$, let

$$P(t,x) = \frac{\partial h}{\partial t}(t,x) = \frac{c_d t}{(t^2 + \|x\|^2)^{(d+1)/2}}$$

$$\text{and } K_j(t,x) = \frac{\partial h}{\partial x_j}(t,x) = \frac{c_d x_j}{(t^2 + \|x\|^2)^{(d+1)/2}},$$

for $1 \le j \le d$. P is the d-dimensional *Poisson kernel* and the functions K_j are the *Riesz kernels*. The constant c_d is chosen so that

$$\int_{\mathbf{R}^d} P(t,x)\,dx = 1 \text{ for } t > 0.$$

For fixed $t > 0$, the functions $K_j(t, \cdot)$ are bounded functions on $\mathbf{R}^d$ which are in $L^p(\mathbf{R}^d)$ for $1 < p < \infty$ (but not in $L^1(\mathbf{R}^d)$).

Since h is harmonic,

$$\overline{D}h(t,x) = P(t,x) - \sum_{j=1}^{d} e_j K_j(t,x)$$

is an augmented Clifford analytic function on H^{d+1}.

Now suppose that $\nu \in M(\mathbf{R}^d)$. We define

$$K^{(j)}(\nu)(t,x) = K_t^{(j)}(\nu)(x) = \int_{\mathbf{R}^d} K_j(t, x - y)\,d\nu(y),$$

for $(t,x) \in H^{d+1}$ and for $1 \le j \le d$. Then $P(\nu) - \sum_{j=1}^{d} e_j K^{(j)} j(\nu)$ is an augmented Clifford analytic function on H^{d+1}. In particular, each $K^{(j)}(\nu)$ is a harmonic function on H^{p+1}.

We can also define $K^{(j)}(f)(t,x)$, for $f \in L^p(\mathbf{R}^d)$, for $1 \le p < \infty$. When $1 < p < \infty$, things continue to work well. If $f \in L^p(\lambda)$ then $K^{(j)}(f) \in H_p$, so that $K_t^{(j)}(f)$ converges in L^p-norm to a function $K_j(f)$ as $t \searrow 0$. $K_j(f)$ is the *jth Riesz transform* of f. The mapping $f \to K_j(f)$ is a bounded linear mapping, and

$$K^{(j)}(f)(t,x) = P(K_j(f))(t,x).$$

When $p = 1$, things do not work so well. We must consider $H_1(\mathbf{R}^d)$ rather than $L^1(\mathbf{R}^d)$. If $f \in H_1(\mathbf{R}^d)$, then $K^{(j)}(f) \in H_1(H^{d+1})$, so that $K_t^{(j)}(f)$ converges in L^1-norm to a function $K_j(f)$ as $t \searrow 0$. Then the jth Riesz transform $K_j(f) \in H^1(\mathbf{R}^d)$.

Nevertheless, the facts that things work well for $1 < p < \infty$, and that $|f|^\alpha$ is subharmonic for $\alpha > (d-1)/d$ if f is an augmented Clifford analytic function, lead to a proof of the d-dimensional version of the theorem of the brothers Riesz.

Theorem 10.7.1 *If ν is a bounded measure on $\mathbf{R}^d$ taking values in $\mathbf{R}^d$, for which $P(\nu)$ is Clifford analytic, then ν is absolutely continuous with respect to Lebesgue measure, and there exists $f \in H^1(\mathbf{R}^d)$ such that $\nu = f\, d\lambda$.*

Proof We shall need the following lemma.

Lemma 10.7.1 *Suppose that u is a continuous subharmonic function on H^{d+1}, and that $u(x+iy) \to 0$ as $y \to 0$ and as $x+iy \to \infty$. Then $u \leq 0$ on H^{d+1}.*

Proof If not, u attains its positive supremum at a point (t_0, x_0) of H^{d+1}. Since u takes values arbitrarily close to zero on the ball

$$U = \{(t,x) : (t-t_0)^2 + \|x - x_0\|^2 < t_0^2\},$$

it follows that

$$\frac{1}{\lambda(U)} \int_U u(t,x)\, d\lambda(t,x) < u(t_0, x_0),$$

contradicting the subharmonicity of u. □

We now turn to the proof of the theorem. Let $u = P(\nu)$. Choose $(d-1)/d < q < 1$ and let $v = \|u\|^q$; v is subharmonic on H^{d+1}, by Theorem 10.6.1. Let $p = 1/q > 1$. Then if $s > 0$ and $v_s(x) = v(s,x)$,

$$\int_{\mathbf{R}^d} |v_s(x)|^p dx = \int_{\mathbf{R}^d} \|u_s(x)\|\, dx \leq \|\nu\|,$$

so that $\|v_s\|_p \leq \|\nu\|^q$.

Let $w_s = P(v_s)$, so that w_s is harmonic in H^{d+1}. The function $d_s(t,x) = v(s+t,x) - w_s(t,x)$ is subharmonic on H^{d+1}. Also $d_s(t,x)) \to v(s,x) - v_s(x) = 0$ as $s \searrow 0$, and $d_s(t,x) \to 0$ as $(t,x) \to \infty$. Thus by Lemma 10.7.1, $v(s+t,x)) \leq w_s(t,x)$, for $(t,x) \in H^{d+1}$.

We now use the fact that if $1 < p < \infty$ then the space L^p is reflexive. This means that if (f_n) is a bounded sequence in L^p then there is a subsequence (f_{n_k}) which converges weakly: there exists $f \in L^p$ such that $\int f_{n_k} g\, d\lambda \to \int fg\, d\lambda$ for each $g \in L^{p'}$ (where $1/p + 1/p' = 1$). Since $\{v_s : s > 0\})$ is bounded in $L^p(\mathbf{R}^d)$, there exists a sequence (s_k), with $s_k \searrow 0$ as $k \to \infty$ such that (v_{s_k}) converges weakly in $L^p(\mathbf{R}^d)$ to V, say. In particular, $w_{s_k}(t,x) = P(v_{s_k})(t,x) \to P(V)(t,x)$ for each

$(t, x) \in H^{d+1}$. Further, $v(s_k + t, x) \to v(t, x)$ as $k \to \infty$. Consequently,

$$\|P(\nu)(t, x)\| = \|u(t, x)\| = v(t, x)^p \leq (P(V)(t, x))^p \leq (P(V)^*(x))^p,$$

and so $P(\nu)^* \in L^1(\mathbf{R}^d)$. Thus $P(\nu) \in H_1(H^{d+1})$, and so there exists $f \in H_1(\mathbf{R}^d)$ such that $P(\nu) = P(f)$. Thus $\nu = f\,d\lambda$. $\qquad\square$

10.8 The Dirac operator on a Riemannian manifold

So far, we have considered the Dirac operator, and its properties, defined on functions on an open subset of Euclidean space. Even more important are the properties of a Dirac operator defined on the sections of a vector bundle over a Riemannian manifold. These properties play a large part in proofs of the Atiyah-Singer index theorem. It is unfortunate that a substantial amount of knowledge of Riemannian geometry is needed in order to give an accurate and detailed account of this; instead we shall give a brief and superficial description of what is involved, if only to encourage the interested reader to consult fuller accounts. To do so, we assume some rudimentary knowledge of differential geometry.

A smooth differential manifold M is a *Riemannian manifold* if there is a smoothly varying inner product, defined on the fibres T_x of the tangent bundle TM of M. Using this, it is possible to define a smooth *Clifford algebra bundle* $\mathcal{A}(TM)$ over M; if $x \in M$ then the fibre $\mathcal{A}_x$ is a universal Clifford algebra for the inner-product space T_x. As usual, we consider T_x as the space of vectors in $\mathcal{A}_x$.

We can now define Dirac bundles over M. A vector bundle S over M is a *Dirac bundle over M* if, first, each fibre S_x is a left $\mathcal{A}_x$-module. Secondly, we require S to be a Riemannian manifold with a canonical torsion-free connection ∇ which satisfies the following compatibility conditions:

(i) if a is a unit vector in T_x and $s \in S_x$ then $\langle a.s, a.s \rangle_x = \langle s, s \rangle_x$. Thus

$$\langle a.s, s \rangle_x = \langle a.s, a^2.s \rangle_x = -\langle s, a.s \rangle_x,$$

so that a defines a skew-symmetric operator on the fibre S_x;

(ii) if a is a smooth cross-section of $\mathcal{A}(TM)$ and s is a smooth cross-section of S, then

$$\nabla(a.s) = (\nabla a).s + a.(\nabla s).$$

As an example, the Clifford algebra bundle $\mathcal{A}(TM)$ is itself a Dirac bundle over M. It is however not always possible to define a corresponding spinor bundle.

The *Dirac operator* D is now defined as a first-order linear differential operator on the vector space $\Gamma(S)$ of smooth sections of the Dirac bundle S. If $x \in M$ then

$$Ds(x) = \sum_{j=1}^{d} e_j \nabla_{e_j} s(x),$$

where $(e_1, \ldots, e_d)$ is an orthogonal basis for the fibre T_x; as in the Euclidean case, this definition does not depend on the choice of orthogonal basis. It then turns out that D^2 is an elliptic essentially self-adjoint operator on $\Gamma(S)$. Further

$$D^2 s = \sum_{i=1}^{d} e_i \nabla_{e_i} \left(\sum_{j=1}^{d} e_j \nabla_{e_j} s \right)$$

$$= -\sum_{i=1}^{d} \nabla_{e_i}^2 s + \sum_{1 \leq i < j \leq d} e_i e_j (\nabla_{e_i} \nabla_{e_j} - \nabla_{e_j} \nabla_{e_i}) s.$$

The operator $\sum_{i=1}^{d} \nabla_{e_i}^2$ is an operator of Laplacian type; the second term is a multiple of the curvature tensor. This means that the Dirac operator involves not only harmonic properties of sections, but also the geometry of the Riemannian manifold, and it is this that makes it a powerful tool.

This is as far as it is appropriate to go here. Suggestions for further reading are made in Chapter 12.

11

Representations of Spin_d and $SO(d)$

11.1 Compact Lie groups and their representations

The groups Spin_d and $SO(d)$ are examples of Lie groups. A *Lie group* G is a group which is also a smooth manifold, and which has the property that the mappings $(g, h) \to gh : G \times G \to G$ and $g \to g^{-1} : G \to G$ are smooth (continuous and infinitely differentiable). The product of finitely many Lie groups is a Lie group; an important example is the torus $\mathbf{T}^k$, which is a connected compact abelian Lie group. In fact, if G is a k-dimensional connected compact abelian Lie group, then G is isomorphic to $\mathbf{T}^k$.

We shall consider (complex) representations of a compact Lie group G. A *representation* π of G is a continuous homomorphism of G into the group $GL(V)$ of invertible linear mappings of a complex vector space V. It is said to be *finite-dimensional* if V is finite-dimensional, and to be a *character* if $V = \mathbf{C}$. We can for example consider the left regular representation π of G in $GL(C(G))$, where $C(G)$ is the Banach space of continuous complex-valued functions on G, defined by

$$(\pi(g)(f)(h) = f(g^{-1}h), \quad \text{for} \ \ f \in C(G) \ \ \text{and} \ \ g, h \in G.$$

This is however an infinite-dimensional representation, and we look for finite-dimensional subspaces which are invariant under this representation.

Two representations $\pi_1 : G \to GL(V_1)$ and $\pi_2 : G \to GL(V_2)$ are said to be *equivalent* if there is a linear isomorphism T of V_1 onto V_2 such that

$$\pi_2(g)(T(v)) = T(\pi_1(g)(v)) \quad \text{for all} \ \ g \in G, v \in V_1.$$

The linear isomorphism T is called an *intertwining operator*.

A finite-dimensional representation is *unitary* if V is an inner-product space, and $\pi(G) \subseteq U(V)$, where $U(V)$ is the group of unitary elements of $GL(V)$. In this case,

$$\langle \pi(g)(v), \pi(g)(v') \rangle = \langle v, v' \rangle \quad \text{for all} \quad g \in G, v, v' \in V.$$

If π is a finite-dimensional representation of the compact group G in $GL(V)$, then there exists an inner product on V with respect to which π is unitary. (This is achieved by putting an arbitrary inner product on V, and averaging, using Haar measure on G.)

Suppose that π is a unitary representation of a compact Lie group G in $U(V)$. A linear subspace W of V is *G-invariant* if $\pi(g)(W) \subseteq W$ for all $g \in G$. If W is G-invariant, then so is $W^\perp$, and the orthogonal projection P_W of V onto W satisfies $P_W \pi(g) = \pi(g) P_W$ for all $g \in G$. The representation is *irreducible* if $\{0\}$ and V are the only G-invariant subspaces of V; π is irreducible if and only if whenever $T \in L(V)$ and $\pi(g)T = T\pi(g)$ for all $g \in G$ then T is a scalar multiple of the identity. If π is a unitary representation of a compact Lie group G then V can be decomposed as a G-invariant orthogonal direct sum $V = V_1 \oplus \ldots \oplus V_k$, such that, if $\pi_j(g)$ is the restriction of $\pi(g)$ to V_j, each of the representations π_j is irreducible. Because of this, we concentrate our attention on irreducible representations.

As we have seen, there is a double cover given by the short exact sequence

$$1 \longrightarrow D_2 \overset{\subseteq}{\longrightarrow} \text{Spin}_d \overset{\alpha}{\to} SO(d) \longrightarrow 1.$$

We shall use this to study the relationship between the irreducible representations of Spin_d and of $SO(d)$. To do this we shall use the following result. Suppose that G is a compact Lie group whose irreducible representations are known, and that $\beta : G \to H$ is a continuous surjective homomorphism of G onto a compact Lie group H, with kernel K. If π is an irreducible representation of H then $\pi \circ \beta$ is an irreducible representation of G; the mapping $\pi \to \pi \circ \beta$ is then a one-one correspondence between the irreducible representations of H and those irreducible representations of G which map K to the identity operator. Thus to determine the irreducible representations of H we need only pick out those irreducible representations of G which map K to the identity operator.

In fact we shall restrict our attention to the cases where $d \leq 4$; to go futher requires too much of the theory of Lie groups to be considered here.

11.2 Representations of $SU(2)$

It is relatively easy to determine the irreducible representations of $SU(2)$. Recall that the mapping

$$(z, w) \to \begin{bmatrix} z & w \\ -\bar{w} & \bar{z} \end{bmatrix}$$

is a homeomorphism of $S^3 = \{(z, w) : |z|^2 + |w|^2 = 1\}$ onto $SU(2)$. Thus the left regular representation of $SU(2)$ is equivalent to the corresponding representation of $SU(2)$ in $GL(C(S^3))$. We shall consider the latter; if $f \in C(S^3)$ and

$$g^{-1} = \begin{bmatrix} a & b \\ -\bar{b} & \bar{a} \end{bmatrix} \in SU(2)$$

then $\pi(g)(f)(z, w) = f(az + bw, -\bar{b}z + \bar{a}w)$.

We now look for finite-dimensional invariant subspaces of $C(S^3)$. For each $n \in \mathbf{Z}$ let P_n be the vector space of homogeneous polynomials of degree n in two complex variables, considered as functions on $\mathbf{C}^2$. Let V_n be the vector space of the restrictions of functions in P_n to S^3. We denote the function $(z, w) \to z^j w^k$ in V_n by $h_{j,k}$. V_n has dimension $n + 1$; $\{1\}$ is a basis for V_0, and $(h_{n,0}, h_{n-1,1}, \ldots, h_{0,n})$ is a basis for V_n, for $n > 0$. If g^{-1} is as above, and $j + k = n$ then

$$\pi(g)h_{j,k}(z, w) = (az + bw)^j(-\bar{b}z + \bar{a}w)^k$$

so that $\pi(g)(h_{j,k}) \in V_n$, and the vector space V_n is G-invariant. Let π_n be the restriction of π to π_n.

Theorem 11.2.1 *Each representation $\pi_n : G \to L(V_n)$ is irreducible.*

Proof The trivial representation V_0 is irreducible. Suppose that $n > 0$. It is enough to show that if $T \in L(V_n)$ and $T\pi_n(g) = \pi_n(g)T$ for all $g \in G$ then T is a multiple of the identity.

First, let μ be irrational, and let $a = e^{i\mu\pi}$, so that all the powers of a, positive and negative, are different. Let

$$g_a = \begin{bmatrix} a & 0 \\ 0 & a^{-1} \end{bmatrix} = \begin{bmatrix} a & 0 \\ 0 & \bar{a} \end{bmatrix}.$$

Then $g_a \in SU(2)$ and if $j + k = n$ then

$$\pi_n(g_a)(h_{j,k})(z, w) = (\bar{a}z)^j(aw)^k = a^{k-j}h_{j,k}(z, w),$$

so that $h_{j,k}$ is an eigenvector of $\pi_n(g_a)$, with eigenvalue a^{k-j}. Thus

$\pi_n(g_a)$ has distinct eigenvalues. Since $T\pi_n(g_a) = \pi_n(g_a)T$, it follows that $T(h_{j,k}) = \lambda_{j,k}h_{j,k}$, for some $\lambda_{j,k} \in \mathbf{C}$.

Next consider the rotation

$$R = \frac{1}{\sqrt{2}}\begin{bmatrix} 1 & -1 \\ 1 & 1 \end{bmatrix} \text{ with inverse } R^{-1} = \frac{1}{\sqrt{2}}\begin{bmatrix} 1 & 1 \\ -1 & 1 \end{bmatrix}.$$

Then $R \in SU(2)$ and

$$\pi_n(R)(h_{n,0})(z,w) = \frac{(z+w)^n}{2^{n/2}} = \frac{1}{2^{n/2}}\sum_{j=0}^n \binom{n}{j}z^j w^{n-j},$$

so that

$$\pi_n(R)(h_{n,0}) = \frac{1}{2^{n/2}}\sum_{j=0}^n \binom{n}{j}h_{j,n-j}$$

and

$$T\pi_n(R)(h_{n,0})) = \frac{1}{2^{n/2}}\sum_{j=0}^n \binom{n}{j}\lambda_{j,n-j}h_{j,n-j}.$$

On the other hand,

$$\pi_n(R)(T(h_{n,0})) = \lambda_{n,0}\pi_n(R)(h_{n,0}) = \frac{1}{2^{n/2}}\sum_{j=0}^n \binom{n}{j}\lambda_{n,0}h_{j,n-j}.$$

Thus $\lambda_{j,n-j} = \lambda_{n,0}$ for all $0 \le j \le n$, and $T = \lambda_{n,0}I$. $\square$

In fact, these are all the irreducible representations of $SU(2)$.

Theorem 11.2.2 *If π is an irreducible representation of $SU(2)$ then π is equivalent to π_n for some $n \in \mathbf{Z}$.*

Proof This requires some basic representation theory. See [BtD], Proposition II.5.3. $\square$

11.3 Representations of Spin$_d$ and $SO(d)$ for $d \le 4$

We now consider the irreducible representations of Spin$_d$ and $SO(d)$ for $d = 2, 3$ and 4.

The groups Spin$_2$ and $SO(2)$ are both abelian, and are isomorphic to $\mathbf{T}$. Thus the irreducible representations of each are the characters, and in each case the dual group is isomorphic to $\mathbf{Z}$. If γ is a character on $SO(2)$, then $\gamma \circ \alpha$ is a character on Spin$_2$. Since $\alpha(e^{te_\Omega})x = e^{2te_\Omega}x$, the mapping $\gamma \to \gamma \circ \alpha$ maps the group $SO(2)'$ of characters on $SO(2)$ onto the group Spin'_2 of characters on Spin_2 whose elements correspond to the even integers in $\mathbf{Z}$.

The group Spin_3 is isomorphic to $SU(2)$, and so we can identify the irreducible representations of Spin_3 with the irreducible representations $\{\pi_n : n \in \mathbf{Z}\}$ of $SU(2)$.

Let us consider this in a little detail. Let us, as in Section 7.5, use the isomorphism $j : \mathcal{A}_{0,3} \to M_2(\mathbf{C})$ defined by the Pauli spin matrices. If we set $f_i = e_i e_\Omega$, so that $j(f_i) = \tau_i$, the associate Pauli matrix, for $1 \leq i \leq 3$, and consider the isomorphism j of $\mathrm{Spin}_{0,3}$ onto $SU(2)$, we find that

$$j(e^{tf_1}) = \begin{bmatrix} \cos t & i\sin t \\ i\sin t & \cos t \end{bmatrix},$$

$$j(e^{tf_2}) = \begin{bmatrix} \cos t & \sin t \\ -\sin t & \cos t \end{bmatrix},$$

$$j(e^{tf_3}) = \begin{bmatrix} e^{it} & 0 \\ 0 & e^{-it} \end{bmatrix}.$$

Let us set $\gamma_n = \pi_n \circ j : \mathrm{Spin}_{0,3} \to GL(V_n)$. Then

$$\gamma_n(e^{tf_1})h_{j,k}(z,w)) = (z\cos t - iw\sin t)^j(-iz\sin t + w\cos t)^k,$$
$$\gamma_n(e^{tf_2})h_{j,k}(z,w)) = (z\cos t - w\sin t)^j(z\sin t + w\cos t)^k,$$
$$\gamma_n(e^{tf_3})h_{j,k}(z,w)) = e^{i(k-j)t}z^j w^k.$$

Now $\gamma_n(-I) = (-1)^n \gamma_n(I)$, so that we obtain an irreducible representation of $SO(3)$ if and only if n is even. If $g \in SO(3)$ and $g = \alpha(\tilde{g})$, let us set $\tilde{\gamma}_{2n}(g) = \gamma_{2n}(\tilde{g})$.

Theorem 11.3.1 *For each even integer $2n$, $\tilde{\gamma}_{2n}$ is an irreducible representation of $SO(3)$. Every irreducible representation π of $SO(3)$ is equivalent to a representation $\tilde{\gamma}_{2n}$, for some even integer $2n$.*

Proof This essentially follows from the remarks at the end of Section 11.1.   □

Further, if $R_{i,t}$ is a rotation in $SO(3)$ about the ith axis through an angle t, then $R_{i,t} = \alpha(e^{tf_i/2})$, for $1 \leq i \leq 3$, and so we can use the formulae above to calculate $\tilde{\gamma}_{2n}(R_{i,t})$.

While we are about it, let us describe the irreducible representations of $O(3)$ and $U(2)$. We need the following theorem.

Theorem 11.3.2 *Suppose that* $\pi_1 : G_1 \to L(V_1)$ *and* $\pi_2 : G_2 \to L(V_2)$ *are irreducible representations. Then the representation* $\pi_1 \otimes \pi_2$ *of* $G_1 \times G_2$ *in* $L(V_1 \otimes V_2)$ *defined by*

$$(\pi_1 \otimes \pi_2)(g_1, g_2)(v_1 \otimes v_2) = \pi_1(g_1)(v_1) \otimes \pi_2(g_2)(v_2)$$

is irreducible. Every irreducible representation of $G_1 \times G_2$ *is equivalent to one of these representations.*

Proof See [BtD] Proposition II.4.14. □

Theorem 11.3.3 *For each even integer* $2n$ *there is a unique irreducible representation* $\sigma_{2n} : O(3) \to L(V_{2n})$ *such that* $\sigma_{2n}(g) = \tilde{\gamma}_{2n}(g)$ *for* $g \in SO(n)$ *and* $\sigma_{2n}(g) = \sigma(-g)$ *for* $g \in O(3) \setminus SO(3)$, *and there is a unique irreducible representation* $\tau_{2n} : O(3) \to L(V_{2n})$ *such that* $\tau_{2n}(g) = \tilde{\gamma}_{2n}(g)$ *for* $g \in SO(n)$ *and* $\tau_{2n}(g) = -\tau(-g)$ *for* $g \in O(3) \setminus SO(3)$, *and every irreducible representation of* $O(3)$ *is equivalent to one of these representations.*

Proof For $O(3)$ is isomorphic to $SO(3) \times D_2$. □

Next, let us consider the irreducible representations of $U(2)$. The mapping $\beta : (e^{i\theta}, g) \to e^{i\theta}g$ is a surjective homomorphism of $\mathbf{T} \times SU(2)$ onto $U(2)$, with kernel $\{(1, I), (-1, -I)\}$. Thus we have a short exact sequence

$$1 \longrightarrow D_2 \overset{\subseteq}{\longrightarrow} \mathbf{T} \times SU(2) \overset{\beta}{\longrightarrow} U(2) \longrightarrow 1.$$

Every irreducible representation of $\mathbf{T} \times SU(2)$ is of the form $\pi_{m,n}(e^{i\theta}, g) = e^{im\theta}\pi_n(g)$, for some $(m, n) \in \mathbf{Z} \times \mathbf{Z}^+$. Since $\pi_{m,n}((-1, -I) = (-I)^{m+n}$, the irreducible representations of $U(2)$ are derived from the representations $\{\pi_{m,n} : m + n \text{ even}\}$. Note also that $\det g = e^{2i\theta}$. If $n = 2k$ is even, we therefore have a sequence $(\tilde{\pi}_{m,2k})_{m \in \mathbf{Z}}$ of irreducible representations defined by $\tilde{\pi}_{m,2k}(g) = (\det g)^m \pi_{2k}(h)$. If $n = 2k + 1$ is odd, then we have a sequence $(\tilde{\pi}_{m,2k+1})_{m \in \mathbf{Z}}$ of irreducible representations defined by $\tilde{\pi}_{m,2k+1}(g) = e^{i\theta}(\det g)^m \pi_{2k+1}(h)$.

Finally, let us consider the irreducible representations of Spin_4 and $O(4)$. The group Spin_4 is isomorphic to $SU(2) \times SU(2)$, and so it follows from Theorem 11.3.2 that every irreducible representation of Spin_4 is isomorphic to a representation $\gamma_m \otimes \gamma_n$, for $(m, n) \in \mathbf{Z}^2$. The tensor product $V_m \otimes V_n$ can be considered as a linear subspace of $C(S^3 \times S^3)$: if $(g, h) \in SU(2) \otimes SU(2)$ and $f \in V_m \otimes V_n \subseteq C(S^3 \times S^3)$, then

$$\gamma_m \otimes \gamma_n(g, h)(f)(s, t) = f(\pi_m(g^{-1})(s), \pi_n(h^{-1})(t)).$$

The representations $\gamma_m \otimes \gamma_n$ and $\gamma_n \otimes \gamma_m$ are equivalent, and so every irreducible representation of Spin$_4$ is equivalent to some $\gamma_m \otimes \gamma_n$, with $m \leq n$.

Since $(\gamma_m \otimes \gamma_n)(-1, -1) = 1$ if and only if $m+n$ is even, it follows that every irreducible representations of $SO(4)$ comes from an irreducible representation $\gamma_m \otimes \gamma_n$ with $m \leq n$ and $m + n$ even.

Exercise

1. Suppose that $0 \leq n_1 \leq m_1$ and $0 \leq n_2 \leq m_2$ and that $m_1 n_1 = m_2 n_2$. Show that the irreducible representations $\pi_{m_1} \otimes \pi_{n_1}$ and $\pi_{m_2} \otimes \pi_{n_2}$ are isomorphic if and only if $m_1 = m_2$, $n_1 = n_2$.

12

Some suggestions for further reading

This book is only an introduction to Clifford algebras. Here are some suggestions for further reading; they are not meant to provide a comprehensive bibliography, but rather to indicate where to go next.

The algebraic environment

Further results about multilinear mappings and tensor products, are given in standard textbooks, such as Cohn [Coh], Jacobson [Jac] and Mac Lane and Birkhoff [MaB]. The results are presented in the more general setting of modules over a commutative ring. This leads to serious problems which do not arise in the vector space case. The idea of considering representations of algebras in terms of modules extends to infinite-dimensional algebras, such as C^*-algebras. A good starting point for this is the book by Lance [Lan].

The proof of the existence of the tensor product of two modules over a commutative ring, as described in the remark at the end of Section 3.2, does not lead to a simple description of the structure of the tensor product, and there are also problems with torsion. Tensor products of vector spaces are so much more straightforward that they deserve to be treated separately.

Quadratic spaces

Lam [Lam] is the standard work on quadratic forms, and is a goldmine of mathematics. It considers quadratic forms over fields not of characteristic 2. Here, extra complications arise; many of these concern the nature of the set F^2 of elements of a field which are squares, and involve interesting problems in number theory. On the other hand, many of the results of Chapter 3 carry over to this more general situation. For example, the proof of the Cartan-Dieudonné theorem (Theorem 4.8.1), which is due to Artin, applies in this more general situation.

Clifford algebras

There are many books on Clifford algebras. The works by Artin [Art] and Chevalley [Che] are now principally of historic interest. As Bourguignon writes in his *Postface*, 'Chevalley's style is as dry and systematic as possible'; Bourguignon's *Postface* is itself well worth reading. Of the books that are concerned with real Clifford algebras, let us first mention Lounesto [Lou]. This is a discursive account; it begins by considering many examples, of Clifford algebras, of spin groups and of spinors, and does not give the definition of a general Clifford algebra until the second half of the book. The examples are interesting; the book is neither dry, nor systematic. The two books by Porteous, [Por1, Por2] , approach Clifford algebras from a geometric viewpoint, and therefore make up for the lack of geometry in the present work. They consider topics not included, or barely touched on in the present book: the classical groups and their relation to Clifford algebras, the Cayley algebra, conformal groups, and triality. This last topic concerns a group of automorphisms of $Spin_8$ of order 3 which does not project down to $SO(8)$. These topics are also studied in the book by Harvey [Har]. This comes in two separate parts; the first is concerned with normed algebras and calibrations, the second with Clifford algebras and spinors. The presentation of Clifford algebras is independent of Part I, but the two parts come together when spinor spaces are equipped with a natural inner product. Harvey suggests that the reader may wish to start the book with Part II. Do so; this is a most enjoyable book. He claims that his book is intended to be a collection of examples; it includes many more calculations than the present book does, but presented in a very different way. The books by Artin and Chevalley are concerned with Clifford algebras over general fields. It is also possible to construct Clifford algebras based on modules with coefficients in a commutative unital algebra; these are studied in the book by Hahn [Hah] and the recent magisterial book by Helmstetter and Micali [HeM].

Clifford algebras can be constructed for infinite-dimensional spaces. Plymen and Robinson [PlR] start with a *real* infinite-dimensional inner-product space, and use it to construct a *complex* Clifford algebra; from this they construct a C^*-Clifford algebra, and a von Neumann Clifford algebra, and study the properties of these algebras. Similar algebras are considered by Meyer [Mey], but are approached from a very different direction; Meyer treats the subject as a branch of non-commutative, or *quantum*, probability theory.

Clifford algebras and harmonic analysis

The book by Brackx, Delange and Sommen [BDS] gives a comprehensive account of Clifford analysis. The most useful starting point, however, is the book by Gilbert and Murray [GiM]. This begins by developing the theory of Clifford algebras, and then considers Dirac operators, Riesz transforms and Clifford analyticity, operators of Dirac type and the representation theory of spin groups (particularly in the Euclidean case) and ends with a brief account of the Atiyah-Singer index theorem. The lecture notes by Alan McIntosh *'Clifford algebras, Fourier series and harmonic functions in a Lipschitz domain'*, in [Rya], also provide a comprehensive introduction to Clifford analysis (although McIntosh works with complex Clifford algebras), and there is much else of interest in these conference proceedings.

The Riesz transforms are discussed in the classic works by Stein [Ste] and Stein and Weiss [StW], but without placing them in the context of Clifford algebras. Good accounts of the background theory of Hardy spaces are given in the books by Garcia-Cuerva and Rubio de Francia [GRF] and by Duoandikoetxea [Duo].

Clifford analysis on Riemannian manifolds

In their proof of the index theorem [AtS], Atiyah and Singer defined Dirac operators on certain Riemannian manifolds. In the process they presented Clifford algebras in a new way, which has been followed ever since. The paper by Atiyah, Bott and Shapiro [ABS] gives an account of this: it is well worth reading. The book by Lawson and Michelsohn [LaM] is the standard work on Clifford analysis on Riemannian manifolds; the authors describe their work as a 'modest introduction', whose purpose is 'to give a leisurely and rounded presentation' of the results of Atiyah and Singer; it is much, much more than this. It does however require a good grounding in the theory of Riemann manifolds.

Subsequent versions of the index theorem relate the Clifford analysis to the asymptotics of the heat kernel on the Riemannian manifold. Short accounts are given by Gilbert and Murray [GiM] and by Roe [Roe]; an encyclopedic account is given by Berline, Getzler and Vergne [BGV]. Although this book includes a preliminary chapter on differential geometry, it makes heavy demands on the reader.

Spin groups and representation theory

Spin groups have long been considered as double covers of classical Lie groups. There is a great difference between the theory of the representations of compact Lie groups and the theory of the representations of more general locally compact Lie groups, and it is well worth starting with the former. The excellent book by Bröcker and tom Dieck [BtD] is the standard text on this, and another good account is given in the recent book by Sepanski [Sep]. Both books explain the use of Clifford algebras and spin groups in representation theory.

Applications to physics

Finally, let us mention some books that link Clifford algebras with physics. The book by Doran and Lasenby [DoL] proceeds at a leisurely pace. Clifford algebras (called here by their alternative name 'geometric algebras') are introduced and studied as fundamental tools for the description of the physical world, rather than as objects in their own right, and much of the attention is concentrated on three-dimensional Euclidean space, and four-dimensional space-time. Nevertheless, a great deal of physical theory is expressed in a convincing way in terms of Clifford algebras. Similar remarks apply to the articles in [Bay]. The two volumes by Penrose and Rindler [PeR] use Clifford algebras to explore space-time in a very geometric manner. Many writers on Clifford algebras confess their lack of knowledge of theoretical physics. I must do the same; I have found the book by Sudbery [Sud] helpful in making connections with the mathematical theory of quantum mechanics.

References

[Art]   E. Artin, *Geometric algebra,* John Wiley, New York, 1988.

[ABS]   M. F. Atiyah, R. Bott and A. Shapiro, Clifford modules, *Topology* **3** (Supplement 1) (1964), 3-38.

[AtS]   M. F. Atiyah and I. M. Singer, The index of elliptic operators on compact manifolds, *Bull. Amer. Math. Soc.* **69** (1963), 422-433.

[Bay]   William E. Baylis (editor), *Clifford (Geometric) Algebras,* Birkhäuser, New York, 1996.

[BGV]   Nicole Berline, Ezra Getzler and Michèle Vergne, *Heat Kernels and Dirac Operators,* Springer, Berlin, 1996.

[BDS]   F. Brackx, R. Delanghe and F. Sommen, *Clifford Analysis,* Pitman, London, 1983.

[BtD]   Theodor Bröcker and Tammo tom Dieck, *Representations of Compact Lie groups,* Springer, Berlin, 1995.

[Che]   Claude Chevalley, *The Algebraic Theory of Spinors and Clifford Algebras,* Collected works, Volume 2, Springer, Berlin, 1997.

[Cli1] W. K. Clifford, Applications of Grassmann's extensive algebra, *Amer. J. Math.* **1** (1876) 350-358.

[Cli2] W. K. Clifford, *On the classification of geometric algebras, Mathematical papers, William Kingdon Clifford,* AMS Chelsea Publishing, 2007, 397-401.

[Coh]   P. M. Cohn, *Classic Algebra,* John Wiley, Chichester, 2000.

[Dir]   P. A. M. Dirac, *The Principles of Quantum Mechanics,* Fourth Edition, Oxford University Press, 1958

[DoL]   Chris Doran and Anthony Lasenby, *Geometric Algebra for Physicists,* Cambridge University Press, Cambridge, 2003.

[Duo]   Javier Duoandikoetxea, *Fourier Analysis,* Amer. Math. Soc., Providence, RI, 2001.

[GRF]   J. Garcia-Cuerva and J. L. Rubio de Francia, *Weighted Norm Inequalities and Related Topics,* North Holland, Amsterdam, 1985.

[GiM]   John E. Gilbert and Margaret M. E. Murray, *Clifford Algebras and Dirac Operators in Harmonic Analysis,* Cambridge University Press, Cambridge, 1991.

[Hah] Alexander J. Hahn, *Quadratic Algebras, Clifford Algebras and Arithmetic Witt Groups,* Springer, Berlin, 1994.

[Har] F. Reese Harvey, *Spinors and Calibrations,* Academic Press, San Diego, CA, 1990.

[HeM] Jacques Helmstetter and Artibano Micali, *Quadratic Mappings and Clifford Algebras,* Birkhaüser, Basel, 2008.

[Jac] Nathan Jacobson, *Basic Algebra I and II,* W.H. Freeman, San Francisco, CA, 1974, 1980

[Lam] T. Y. Lam, *Introduction to Quadratic Forms over Fields,* Amer. Math. Soc., Providence, RI, 2004.

[Lan] E. C. Lance *Hilbert C^*-modules,* London Mathematical Society Lecture notes **210**, Cambridge University Press, Cambridge, 1995.

[LaM] H. Blaine Lawson, Jr. and Marie-Louise Michelsohn, *Spin Geometry,* Princeton University Press, Princeton, NJ, 1989.

[Lou] Pertti Lounesto, *Clifford Algebras and Spinors,* London Mathematical Society Lecture notes **286**, Cambridge University Press, Cambridge, 2001.

[MaB] Saunders Mac Lane and Garrett Birkhoff, *Algebra* Third Edition, AMS Chelsea, 1999.

[Mey] Paul-André Meyer, *Quantum Probability for Probabilists,* Springer Lecture Notes in Mathematics **1538**, Springer, Berlin, 1993.

[PeR] R. Penrose and W. Rindler, *Spinors and Space-time I and II,* Cambridge University Press, Cambridge, 1984, 1986.

[PlR] R. J. Plymen and P. L. Robinson, *Spinors in Hilbert Space,* Cambridge University Press, Cambridge, 1994.

[Por1] I. R. Porteous, *Topological Geometry,* Cambridge University Press, Cambridge, 1981.

[Por2] I. R. Porteous, *Clifford Algebras and the Classical Groups,* Cambridge University Press, Cambridge, 1995.

[Roe] John Roe, *Elliptic Operators, Topology and Asymptotic Methods,* Second edition, Chapman and Hall/CRC, London 2001.

[Rya] John Ryan (editor), *Clifford Algebras in Analysis and Related Topics,* CRC Press, Boca Raton, FL, 1996.

[Sep] Mark R. Sepanski, *Compact Lie Groups,* Springer, Berlin, 2007.

[Ste] Elias M. Stein, *Singular Integrals and Differentiability Properties of Functions,* Princeton University Press, Princeton, NJ, 1970.

[StW] Elias M. Stein and Guido Weiss, *Introduction to Fourier Analysis on Euclidean Spaces,* Princeton University Press, Princeton, NJ, 1971.

[Sud] Anthony Sudbery, *Quantum Mechanics and the Particles of Nature,* Cambridge University Press, Cambridge, 1986.

Glossary

Ad'_g Adjoint conjugation (137)

$\mathcal{A}(E,q)$, $\mathcal{A}_{p,m}$, $\mathcal{A}_d$ Universal Clifford algebras (88,90)

$\mathcal{A}^+(E,q)$, $\mathcal{A}^+_{p,m}$, $\mathcal{A}^+_d$ The even subalgebra of a universal Clifford algebra (92)

$\mathcal{A}^-(E,q)$, $\mathcal{A}^-_{p,m}$, $\mathcal{A}^-_d$ The odd part of a universal Clifford algebra (92)

$A^k(E,F)$, $A^k(E)$ Spaces of alternating linear mappings and forms (45)

A_n The group of even permutations of $\{1,\ldots,n\}$ (9)

$An(E,q)$ The anisotropic elements of (E,q) (69)

A^{opp} The opposite algebra (18)

$\mathrm{Aut}(G)$ The group of automorphisms of the group G (8)

$\boldsymbol{B}$, $\tilde{\boldsymbol{B}}$ The magnetic field (159)

curl The curl operator (166)

$\mathcal{B}(E,q)$, $\mathcal{B}_{p,m}$, $\mathcal{B}_d$ Non-universal Clifford algebras (99)

$B(E_1,E_2;F)$, $B(E_1,E_2)$, $B(E)$ Spaces of bilinear mappings and forms (37)

$\mathbf{C}$, $\mathbf{C}^*$ The complex numbers, the non-zero complex numbers (1,10)

$C_G(A)$, $C_A(B)$ The centralizer of a set A in a group G, of a set B in an algebra A (9,18)

$C(U,F)$, $C^k(U,F)$ The space of continuous, k times continuously differentiable, functions on U with values in F (156)

$\mathcal{D}$, $\overline{\mathcal{D}}$ The augmented Dirac operator and its conjugate (170)

d_ϕ An annihilation operator (47)

D, D_{2n} The full dihedral group, the dihedral group of order $2n$ (10)

D_2 The multiplicative group $\{1,-1\}$ (9)

D_q The standard Dirac operator (157)

det The determinant (18, 45)

$\dim E$ The dimension of a vector space E (12)

e^a The exponential function (20)

e_Ω A volume element (96)

$\boldsymbol{E}$, $\tilde{\boldsymbol{E}}$ The electric field (159)

(E,q) A quadratic space (61)

$\mathrm{End}(A)$ The algebra of unital automorphisms of a unital algebra (18)

$\mathrm{End}_A(M)$ The algebra of A-automorphisms of an A-module M (29)

$G(A)$ The group of invertible elements of a unital algebra (19)

$\mathcal{G}(E,q)$ The group of invertible elements of $\mathcal{A}(E,q)$ (101)

$GL(E)$, $GL_d(K)$ The general linear group of operators, of matrices (19)

$\mathrm{Gp}(A)$, $\mathrm{Gp}(g)$ The group generated by A, by $\{g\}$ (7)

H, **H***, **H**$_1$ The quaternions, non-zero quaternions, quaternions of norm 1 (25,27)

H_{2p} Standard hyperbolic space of dimension $2p$ (67)

$H_d(\mathbf{C})$ The space of $d \times d$ Hermitian matrices (146)

H^{d+1} The upper half-space (167)

$H(f)$ The Hilbert transform of f (170)

$H_p(H^{d+1})$ A Hardy space (169)

$\mathrm{Hom}_A(M_1, M_2)$ The space of A-module homomorphisms from M_1 to M_2 (29)

$HU(2)$ The hyper-unitary, or symplectic, group (147)

$\tilde{\imath}$ The canonical inclusion (94)

$Iso(E, q)$ The isotropic elements of (E, q) (69)

J A matrix representing a quarter-turn rotation of the plane (23)

J, $\tilde{\mathbf{J}}$ The current density (159)

K_j The jth Riesz kernel (175)

$L(E, F)$, $L(E)$ The vector space of linear mappings of E into F, of E into itself (14)

$L_h(H)$ The space of Hermitian operators on a Hilbert space H (155)

m_x A creation operator (47)

$M(E_1, \ldots, E_k; F)$, $M^k(E; F)$, $M(E, 1 \ldots, E_k)$, $M^k(E)$ Spaces of multilinear mappings and forms (36)

$M(\mathbf{R}^d)$ The space of signed Borel measures on $\mathbf{R}^d$ (168)

$M_{m,n}$, $M_d(A)$ The vector space of $m \times n$ matrices, of $d \times d$ matrices with values in an algebra (13,17)

$N(T)$, $n(T)$ The nullspace of a linear mapping, its nullity (13)

$o(G)$, $o(g)$ The order of the group g, the element g (8)

$\mathcal{N}$, $\mathcal{N}_*$, $\mathcal{N}_{\pm 1}$, $\mathcal{N}_1$ Subgroups of $\mathcal{G}(E, q)$ (102)

$\mathcal{N}^+$, $\mathcal{N}_*^+$, $\mathcal{N}_{\pm 1}^+$, $\mathcal{N}_1^+$ Subgroups of $\mathcal{G}(E, q) \cap \mathcal{A}^+(E, q)$ (102)

$O(E, q)$, $O(p, m)$, $O(d)$ The orthogonal group (71)

(p, m) The signature of a quadratic form (65)

$\mathrm{Pin}(E, q)$ The Pin group (139)

$Pu(\mathbf{H})$, $Pu(A)$ The pure quaternions, the pure elements of an algebra A (25, 43)

$\mathcal{Q}$ The quaternionic group (11)

Q A matrix reprenting a reflection in the plane, or the quadratic form on a universal Clifford algebra (23)

$\mathrm{rank}(T)$ The rank of T (13)

R, **R*** The real numbers, the non-zero real numbers (9)

$\mathbf{R}_{p,m}$, $\mathbf{R}_d$ Standard regular quadratic space, standard Euclidean space (67)

R_θ Rotation of $\mathbf{R}^2$ through an angle θ (74)

$\mathrm{span}(A)$ The span of A [12]

$S^k(E, F)$, $S^k(E)$ Spaces of symmetric k-linear mappings and forms (49)

$S_{p,m}$ A spinor space (115)

$SO(E, q)$, $SO(p, m)$, $SO(d)$ The special orthogonal group (72)

$\mathrm{Spin}(E, q)$ The Spin group (139)

$\mathrm{Spin}^+(E, q)$, $\mathrm{Spin}^-(E, q)$ Elements of $\mathrm{Spin}(E, q)$ with quadratic norm 1, -1, respectively (139)

$SU(d)$ The special unitary group (84)

T The group $\{z \in \mathbf{C} : |z| = 1\}$ (10)

TM The tangent bundle (177)

U A matrix representing a reflection in the plane (23)

$U(E, \langle ., . \rangle)$, $U(d)$ The unitary group (84)
$w = \min(p, m)$ The Witt index of a quadratic form (66)
Z The integers (9)
Z$_2$ The additive group $\{0, 1\} = \mathbf{Z}/2\mathbf{Z}$ (9)
$Z(G)$, $Z(A)$ The centre of a group G, of an algebra A (9)

$\alpha(g)$ Adjoint conjugation (137)
$\gamma_0, \ldots, \gamma_3$ The Dirac matrices (126)
$\Gamma_{3,1}$ The Dirac algebra (126)
$\Gamma(E, q)$ The Clifford group of (E, q) (137)
$\Gamma(S)$ The space of smooth sections of a bundle S (178)
Δ, Δ_q The Laplacian (157)
Δ The quadratic norm function (26, 101)
ρ_x A simple reflection in the direction x (76)
$\sigma_0, \ldots, \sigma_3$ The Pauli spin matrices (24)
Σ_X, Σ_n The group of permutations of X, of $\{1, \ldots, n\}$ (9)
τ, τ_n The trace, the normalized trace (43)
$\tau_0, \ldots, \tau_3$ The associate Pauli matrices (24)
$\Omega = \Omega_d$ The set $\{1, \ldots, d\}$ (89)

∇ A connection (177)
$a \to a'$, $a \to a^*$, $a \to \bar{a}$ The principal automorphism, the principal anti-automorphism, conjugation, in a universal Clifford algebra (91-93)
$(a_1 \otimes b_1)_g(a_2 \otimes b_2)$ The graded tensor product of $a_1 \otimes b_1$ and $a_2 \otimes b_2$ (56)
$A \otimes_g B$ The super-algebra product of A and B (56)
$A^\perp$ The annihilator of A, or the set orthogonal to A (15)
$[C]_A$, $[c_1, \ldots, c_n]_A$ The A-submodule generated by C, by $\{c_1, \ldots, c_n\}$ (29)
E', E'' The dual, the bidual of E (14)
$E_1 \otimes \cdots \otimes E_k$ The tensor product of $E_1, \ldots, E_k$ (38)
$\otimes^*(E)$ The tensor algebra of E (40)
$\wedge^k(E)$ The k-th exterior product of E (45)
$\bigwedge^*(E)$, $\bigwedge^+(E)$, $\bigwedge^-(E)$ The exterior algebra of E, and its odd and even parts (46)
g^* The maximal function of g (169)
$T_1 \otimes \cdots \otimes T_k$ The tensor product of linear operators $T_1, \ldots, T_k$ (41)
T' The transposed mapping (14)
T^a, (t_{ij}^a) The adjoint mapping, the adjoint matrix (68)
T^* The adjoint mapping, in a complex inner-product space (84)
$x_1 \otimes \cdots \otimes x_k$ An elementary tensor (38)
$x_1 \otimes_s \cdots \otimes_s x_k$ An elementary symmetric tensor (49)
$\otimes_s^k(E)$ The space of symmetric k-linear forms on E (49)
$\bigotimes_s^k(E)$, $\bigotimes_s^*(E)$ The kth symmetric tensor product of E, the symmetric tensor algebra of E (50)
$x_1 \wedge \ldots \wedge x_k$ The alternating or wedge product of $x_1, \ldots, x_k$ (45)
$\langle x, y \rangle$ The inner product of x and y (62)

Index

Printed in the United States
By Bookmasters